Nuclear Energy and Global Governance

The book considers the implications of the nuclear energy revival for global governance in the areas of safety, security and non-proliferation.

Increased global warming, the energy demands of China, India and other emerging economic powerhouses and the problems facing traditional and alternative energy sources have led many to suggest that there will soon be a nuclear energy 'renaissance'. This book examines comprehensively the drivers of and constraints on the revival, its nature and scope and the possibility that nuclear power will spread significantly beyond the countries which currently rely on it. Of special interest are developing countries which aspire to have nuclear energy and which currently lack the infrastructure, experience and regulatory structures to successfully manage such a major industrial enterprise. Of even greater interest are countries that may see in a nuclear energy programme a 'hedging' strategy for a future nuclear weapons option.

Following on from this assessment, the author examines the likely impact of various revival scenarios on the current global governance of nuclear energy, notably the treaties, international organizations, arrangements and practices designed to ensure that nuclear power is safe, secure and does not contribute to the proliferation of nuclear weapons. The book concludes with recommendations to the international community on how to strengthen global governance in order to manage the nuclear energy revival prudently.

This book will be of much interest to students of energy security, global governance, security studies and IR in general.

Trevor Findlay holds the William and Jeanie Barton Chair in International Affairs at the Norman Paterson School of International Affairs, Carlton University in Ottawa, Canada. He is also the Director of the Canadian Centre for Treaty Compliance.

Routledge Global Security Studies

Series editors: Aaron Karp, Regina Karp and Terry Teriff

Nuclear Energy and Global Governance

Ensuring safety, security and non-proliferation

Trevor Findlay

Routledge
Taylor & Francis Group

LONDON AND NEW YORK

First published 2011
by Routledge
2 Park Square, Milton Park, Abingdon, Oxon OX14 4RN

Simultaneously published in the USA and Canada
by Routledge
711 Third Avenue, New York, NY 10017

Routledge is an imprint of the Taylor & Francis Group, an informa business

First issued in paperback 2012

© 2011 Trevor Findlay

The right of Trevor Findlay to be identified as author of this work has been
asserted by him in accordance with sections 77 and 78 of the Copyright,
Designs and Patents Act 1988.

Typeset in Times by Wearset Ltd, Boldon, Tyne and Wear

British Library Cataloguing in Publication Data
A catalogue record for this book is available from the British Library

Library of Congress Cataloging-in-Publication Data
A catalog record has been requested for this book

ISBN13: 978-0-415-49364-2 (hbk)
ISBN13: 978-0-203-83450-3 (ebk)
ISBN13: 978-0-415-53248-8 (pbk)

Contents

Illustrations

Foreword

Nuclear issues once again have a pivotal place on the international agenda. In April 2010 a new START Agreement was signed by Russia and the United States that will further cut their strategic nuclear arsenals and renew their bilateral verification arrangements. In the same month US President Barack Obama hosted a special summit on nuclear security that agreed that securing nuclear weapons, facilities and materials is a priority task in preventing nuclear terrorism. A follow-up meeting to check progress will be hosted by South Korea in 2012. In May 2010 the parties to the Nuclear Non-proliferation Treaty successfully agreed a final document, after failure five years earlier, that signals continuing strong multilateral support for the non-proliferation regime. At the Group of Eight summit in Muskoka, Canada, in June, nuclear issues occupied a prominent place on the agenda, including extension of the Global Partnership Against the Spread of Weapons and Materials of Mass Destruction. The International Commission on Nuclear Non-proliferation and Disarmament, for which I served as an advisor, has, meanwhile, elaborated a comprehensive agenda for 'eliminating nuclear threats', including moving towards nuclear disarmament.

Less heartening nuclear developments also make this era pivotal. Both Iran and North Korea continue to defy the international community by refusing to comply with their undertakings to verifiably abjure nuclear weapons. Concerns continue to arise about the extent of nuclear smuggling and the possibility that terrorists might acquire and use radiological or nuclear devices. There remains a deep divide in the nuclear non-proliferation regime between states which have given up nuclear weapons and those which retain them. Three nuclear-armed states, India, Israel and Pakistan, still remain outside the regime. The Comprehensive Nuclear Test Ban Treaty has still not entered into force, while negotiations on a fissile material cut-off treaty have not yet even commenced.

It is in this fraught context that there is increasing speculation about a major, worldwide increase – a 'renaissance' – in the use of nuclear energy for generating electricity, driven by climate change, growing electricity demand and the search for energy security. Yet nuclear energy remains controversial on many grounds: among them cost, safety, security, the nuclear waste question and nuclear weapons proliferation.

In tackling these issues and more, Trevor Findlay's book draws substantially on the research carried out as part of the Nuclear Energy Futures project that I had the pleasure of chairing for the Centre for International Governance Innovation. It reinforces the conclusions reached by that project about both the future of nuclear energy and the implications for nuclear global governance in the areas of safety, security and non-proliferation. Paramount among those conclusions was that, based on a careful calculation of drivers and constraints, the nuclear energy 'renaissance' is likely to be slower and less extensive, at least to 2030, than many anticipate. This is good news for global nuclear governance in that it gives the international community a window of opportunity to deal with the gaps and deficits that are apparent in the myriad treaties, institutions and other arrangements that comprise the nuclear regimes. Action to bolster the regimes is particularly necessary if large numbers of states with no previous experience of nuclear energy are to become owners and operators of nuclear power plants. The International Atomic Energy Agency, the paramount global governance body in the nuclear arena, needs particular attention if it is to be ready for a nuclear energy revival.

With the challenge of nuclear proliferation and nuclear security high on the international agenda, not to mention climate change and energy security which provide the context for increased interest in nuclear energy, Trevor Findlay's book could not be more timely. I commend it to the attention of policy-makers and scholars and students alike.

Louise Fréchette

Louise Fréchette is former Deputy Secretary-General of the United Nations and a Distinguished Fellow at the Centre for International Governance Innovation in Waterloo, Ontario, Canada.

Acknowledgements

This work draws substantially on research conducted from 2006 to 20010 as part of the Nuclear Energy Futures (NEF) Project, a collaborative effort by the Centre for International Governance Innovation (CIGI) in Waterloo, Ontario, Canada, and the Canadian Centre for Treaty Compliance (CCTC) at the Norman Paterson School of International Affairs (NPSIA) at Carleton University in Ottawa, Canada. I am indebted to CIGI and its then director John English for appointing me to run the project and to its financial backers for funding the research so generously. I am especially thankful to Louise Fréchette, chair of the project, who first proposed the idea and provided wise policy guidance and furnished high-level support throughout its lifetime.

This book has, naturally, benefitted significantly from the wealth of information, analysis and opinion that emerged from the conferences and workshops, interviews and consultations, and all of the other activities that culminated in the final project report in April 2010. I am particularly grateful to officials of the International Atomic Energy Agency (IAEA) in Vienna for their valuable insights. All of those involved in contributing to the NEF project, including at CIGI and Carleton University, have been acknowledged in the report, which is available at www.cigionline.org.

Of all of the intellectual contributions made to the project which, in turn, have inspired and enabled me to write this book, I am most cognizant of the published NEF papers authored by Justin Alger, John Cadham, Ian Davis, Kenneth Dormuth, David Jackson, Nathaniel Lowbeer-Lewis, David McLellan, Miles Pomper, M. V. Ramana, Aaron Shull and Sharon Squassoni. This book would also not have been as illuminating without being able to draw on the Survey of Emerging Nuclear Energy States (SENES), a database that tracks the progress of countries seeking civilian nuclear energy for the first time and which is a continuing feature of CIGI's website. Several NPSIA Masters students at the CCTC have contributed to SENES at various stages, including Justin Alger, Paul Davis, Amy Fallis, Ray Froklage, Derek de Jong, Jonathan Miller and Alex Sales, along with CIGI's Som Tsoi. For the original graphics and charts contained in this book, most of which were prepared for the NEF report, I acknowledge the work of Derek de Jong, Justin Alger and CIGI graphics designer Steve Cross.

Above all I am grateful to Justin Alger, my former student, principal researcher for the NEF project and Administrator/Researcher at the CCTC. As in all good collaborations between professor and student I have learned a great deal from him. His Masters thesis, 'Nuclear Alibi: the Nuclear Revival and Proliferation' (2009), in addition to dovetailing perfectly with the NEF project, was a significant resource in the writing of the non-proliferation section of this book. In addition to pursuing innumerable research leads and questions, Justin also took on the valiant task of sorting and checking the references, as well as finalizing the charts and graphics. Not least of all he was a constant source of youthful enthusiasm and encouragement.

Finally, I would like to thank Routledge's Andrew Humphrys, who cajoled a book out of me on this subject, and Rebecca Brennan, who expertly steered it to publication.

Acronyms

ABACC	Argentine-Brazilian Agency for Accounting and Control
ABWR	Advanced Boiling Water Reactor
ACR	Advanced CANDU Reactor
ADB	Asian Development Bank
AECL	Atomic Energy of Canada Limited
AFCONE	African Commission on Nuclear Energy
AFNI	L'Agence France Nucléaire International (France)
ALARA	as low as reasonably achievable
ANSN	Asian Nuclear Safety Network
ANWFZ	African Nuclear Weapon-Free Zone Treaty
AP	Additional Protocol (IAEA)
ASE	Atomstroyexport (Russia)
ASME	American Society of Mechanical Engineers
ASN	Nuclear Safety Authority (France)
AU	African Union
BADEA	Arab Bank for Economic Development in Africa
BMWG	Border Monitoring Working Group (IAEA)
BNFL	British Nuclear Fuels Limited
BOG	Board of Governors (IAEA)
BWR	boiling water reactor
CACNARE	Convention on Assistance in the Case of a Nuclear Accident or Radiological Emergency
CANDU	Canada Deuterium Uranium reactor
CBO	Congressional Budget Office (US)
CCGT	combined cycle gas turbine
CCS	carbon capture and storage
CD	Conference on Disarmament (UN)
CDM	clean development mechanism
CENNA	Convention on Early Notification of a Nuclear Accident
CHP	combined heat and power
CIA	Central Intelligence Agency (US)
CISAC	Committee on International Security and Arms Control
CNRA	Committee on Nuclear Regulatory Activities (OECD/NEA)

CNS	Convention on Nuclear Safety
CNSC	Canadian Nuclear Safety Commission (Canada)
CORDEL	Working Group on Cooperation in Reactor Design Evaluation and Licensing (WNA)
CPPNM	Convention on the Physical Protection of Nuclear Material
CSA	Comprehensive Safeguards Agreement (IAEA)
CSS	Commission on Safety Standards (IAEA)
CTBT	Comprehensive Nuclear Test Ban Treaty
CTR	Cooperative Threat Reduction
DOE	Department of Energy (US)
DTI	Department of Trade and Industry (UK)
EBRD	European Bank for Reconstruction and Development (EC)
EC	European Commission
EDF	Electricité de France
EIA	Energy Information Agency (DOE)
ENAC	Early Notification and Assistance Conventions
ENEN	European Nuclear Education Network
ENSREG	European Nuclear Safety Regulators Group
EPAct	US Energy Policy Act (2005)
EPR	Evolutionary Power Reactor (formerly European Power Reactor)
EPREV	Emergency Preparedness Review Teams (IAEA)
EPRI	Electric Power Research Institute
ERNM	Emergency Response Network Manual
Euratom	European Atomic Energy Community (EC)
FAO	Food and Agricultural Organization of the United Nations
FINAS	Fuel Incident Notification and Analysis System
FMCT	Fissile Material Cut-Off Treaty
FMT	Fissile Material Treaty
FOAK	first-of-a-kind
G8	Group of Eight
GAO	Government Accountability Office (US)
GCC	Gulf Cooperation Council
GDP	gross domestic product
GHG	greenhouse gases
GIF	Generation IV International Forum
GNEP	Global Nuclear Energy Partnership
GTCC	gas turbine combined cycle
HEU	highly-enriched uranium
IACRNA	Inter-Agency Committee on Response to Nuclear Accidents
IAEA	International Atomic Energy Agency
IATA	International Air Transport Association
ICAO	International Civil Aviation Organization
ICJ	International Court of Justice
ICNND	International Commission on Nuclear Nonproliferation and Disarmament

ICRP	International Commission on Radiological Protection
ICSANT	International Convention for the Suppression of Acts of Nuclear Terrorism
IDB	Inter-American Development Bank
IEA	International Energy Agency (OECD)
IMO	International Maritime Organization
INIR	Integrated Nuclear Infrastructure Review (IAEA)
INLEX	International Expert Group on Nuclear Liability
INMM	Institute of Nuclear Materials Management
INPO	Institute of Nuclear Power Operations (US)
INPRO	International Project on Innovative Nuclear Reactors and Fuel Cycles
INRA	International Nuclear Regulators Association
INSAG	International Nuclear Safety Group (IAEA)
INSServ	International Nuclear Security Advisory Service (IAEA)
INSSP	Integrated Nuclear Security Support Plan (IAEA)
INTERPOL	International Criminal Police Organization
IPCC	Intergovernmental Panel on Climate Change
IPFM	International Panel on Fissile Materials
IPPAS	International Physical Protection Advisory Service (IAEA)
IRRS	Integrated Regulatory Review Service
IRS	Incident Reporting System (IAEA/NEA)
IsDB	Islamic Development Bank
ISIS	Institute for Science and International Security
ISSAS	International SSAC Advisory Service (IAEA)
ISSC	International Seismic Safety Centre
ITDB	Illicit Trafficking Database (IAEA)
ITE	International Team of Experts (IAEA)
ITER	International Thermonuclear Experimental Reactor
JREMPIO	Joint Radiation Emergency Management Plan of the International Organizations
JSW	Japan Steel Works
KEPCO	Korea Electric Power Corporation
LEU	low enriched uranium
LNG	Liquid Natural Gas
LWGR	light water-cooled graphite-moderated reactor
LWR	light water reactor
MCIF	Major Capital Investment Fund (IAEA)
MDEP	Multinational Design Evaluation Program
MIT	Massachusetts Institute of Technology
MOI	Ministry of Industry (Vietnam)
MOST	Ministry of Science and Technology (Vietnam)
MOX	mixed oxide fuel
NASA	National Aeronautics and Space Administration (US)
NCACG	National Competent Authorities' Coordinating Group

NEA	Nuclear Energy Agency (OECD)
NEI	Nuclear Energy Institute
NEPIO	Nuclear Energy Programme Implementing Organization
NERC	North American Electric Reliability Corporation
NERS	Network of Regulators of Countries with Small Nuclear Programmes
NESA	Nuclear Energy System Assessment
NEWS	Nuclear Events Web-based System
NGO	non-governmental organization
NGSI	Next Generation Safeguards Initiative
NIA	Nuclear Industry Association (UK)
NIF	National Ignition Facility (US)
NII	Nuclear Installations Inspectorate (UK)
NNWS	non-nuclear weapon state (NPT)
NPT	Nuclear Non-proliferation Treaty
NRC	Nuclear Regulatory Commission (US)
NSEL	Nuclear Security Equipment Laboratory (IAEA)
NSF	Nuclear Security Fund (IAEA)
NSG	Nuclear Suppliers Group
NSSG	Nuclear Safety and Security Group (IAEA)
NTI	Nuclear Threat Initiative
NTM	National Technical Means
NUSS	Nuclear Safety Standards (IAEA)
NWFZ	nuclear-weapon-free zone
NWMO	Nuclear Waste Management Organization (Canada)
NWPA	US Nuclear Waste Policy Act (1982)
NWS	nuclear weapon state (NPT)
O&M	operation and maintenance
OECD	Organisation for Economic Cooperation and Development
OEF	operating experience feedback
OER	Operating Experience Reports
OSART	Operational Safety Review Teams (IAEA)
PHWR	pressurized heavy water reactor
POC	Point of Contact
PRIS	Power Reactor Information System
PROSPER	Peer Review of the effectiveness of the Operational Safety Performance Experience Review
PSI	Proliferation Security Initiative
PSR	Periodic Safety Review
PUREX	Plutonium Uranium Extraction
PWR	pressurized water reactor
RADWASS	Radioactive Waste Safety Standards (IAEA)
RANET	Response Assistance Network
RBMK	Reaktor Bolshoy Moshchnosti Kanalniy (High Power Channel-Type Reactor) (Russia)

RDD   radiological dispersal device
REPLIE   Response Plan for Incidents and Emergencies (IAEA)
RWC   Radiological Weapons Convention
SAGSI   Standing Advisory Group on Safeguards Implementation (IAEA)
SAGSTRAM   Standing Advisory Group on the Safe Transport of Radioactive Materials (IAEA)
SAL   Safeguards Analytical Laboratory (IAEA)
SEDO   Safety Evaluation During Operation of Fuel Cycle Facilities (IAEA)
SENES   Survey of Emerging Nuclear Energy States
SMR   small- and medium-sized reactor
SOER   Significant Operating Experience Reports
SOLAS   International Convention for the Safety of Life at Sea
SQP   Small Quantities Protocol (IAEA)
SSAC   State System of Accounting and Control
STUK   Säteilyturvakeskus (Radiation and Nuclear Safety Authority, Finland)
TTA   Nuclear Trade and Technology Analysis unit (IAEA)
UAE   United Arab Emirates
UNFCCC   United Nations Framework Convention on Climate Change
UNSCEAR   United Nations Scientific Committee on the Effects of Atomic Radiation
USSPC   ultra-supercritical pulverized coal
VARANSAC   Vietnam Agency for Radiation Protection and Nuclear Safety Control
VVER   Vodo-Vodyanoi Energetichesky Reactor (Russia)
WANO   World Association of Nuclear Operators
WENRA   Western European Nuclear Regulators Association
WGRNR   Working Group on Regulation of New Reactors (CNRA)
WHO   World Health Organization
WINS   World Institute of Nuclear Security
WMD   weapons of mass destruction
WMO   World Meteorological Organization
WNA   World Nuclear Association
WNTI   World Nuclear Transport Institute
WNU   World Nuclear University (WNA)

Introduction

The first decade of this millennium has seen a global revival of interest in the use of nuclear energy for generating electricity. Since around 2000 a widely shared view has emerged that the coming years will witness an explosion in the use of nuclear energy amounting to a 'renaissance'.[1] The nuclear industry, in the doldrums since the 1979 Three Mile Island accident and the 1986 Chernobyl disaster, has sensed a 'second coming'. Governments and international organizations have expressed growing enthusiasm for a power source that appears relatively 'green', can help cut greenhouse gas emissions and provide greater energy diversity and security. The nuclear trade press, as well as the general media, has touted the revival, often unquestioningly. Commentary in the media has fed on itself, creating hyperbole reminiscent of the nuclear hucksterism of the Atoms for Peace era of the 1960s and 1970s. Journalistic books and academic tomes alike have reinforced the perceived inevitability of a nuclear revival.[2] There has certainly been, then, a revival of interest in nuclear energy. The first question this study considers is whether this interest is likely to be translated into action.

This book will not use the term 'renaissance' but the more neutral 'revival', except when referring to others' characterization of the revival as a renaissance. 'Renaissance' has a romantic air about it that trivializes the serious dilemmas facing governments and others over energy and climate change policies that may well determine the fate of the planet. It also glosses over complicated issues like the still unsolved problem of long-lived, high-level nuclear waste. This study seeks to be disinterested. It is neither pro- nor anti-nuclear, but is an attempt to predict the likely course of nuclear energy worldwide in the coming decades in all its complexities.

The future growth of nuclear energy globally will ultimately be dependent on the confluence of decisions by governments, electricity utilities, the nuclear industry, private and institutional investors and international organizations. Perhaps most important of all will be the decisions of interested stakeholders outside the industry itself, notably the general public as expressed through elections, opinion polls or other means, and the activities of civil society in supporting or opposing nuclear energy.

One way of considering how policy-makers might reach their decisions is to consider the balance of drivers and constraints. The first three chapters of this book are thus devoted to analysing the drivers and constraints that are most likely to influence decision-making about nuclear energy in the coming decades and seeking to discover where the balance is likely to lie. Facts and figures are useful in this effort but ultimately qualitative and other types of judgements, especially about governments' likely policy preferences, are necessary.

Comparisons are made in this work between nuclear power and other sources of electricity generation, including nuclear's traditional competitors, coal and natural gas, but also emerging alternative energy sources. Inevitably, in a work focused on nuclear power, it is impossible to give equal attention to all of its competitors, especially given the rapid developments occurring in generation by wind, solar energy and renewables. Otherwise the book would have been twice the size.

In similar vein, the focus of this study is necessarily on possible developments to 2030, given that prognostications beyond then become ludicrously difficult to make in a rapidly evolving energy and climate change context. Particular attention is, however, paid to aspirant states, especially those in the developing world, that are seeking to acquire nuclear energy for the first time – since it is often these countries that feature in the more extravagant predictions of a global nuclear energy surge.

The second question that this study seeks to answer is the likely impact of a nuclear energy revival on global nuclear governance. The three key areas of global nuclear governance cover safety, security and the non-proliferation of nuclear weapons. Nuclear safety is concerned with preventing harm to people and the environment from the use of nuclear power, whether from nuclear reactors, nuclear materials in situ or in transit, or associated nuclear fuel cycle facilities such as enrichment and reprocessing plants. This has been an enduring concern of the international community, especially since the Chernobyl disaster. Nuclear security, on the other hand, is concerned with preventing attacks on or illicit access to nuclear facilities and materials. This has been the subject of heightened concern since the terrorist attacks of 11 September 2001 on the United States and the resulting fear that terrorists will seek nuclear materials for nuclear weapons or radiological dispersal devices (RDDs), otherwise known as radiological weapons or in popular parlance 'dirty bombs'. A related concern is that terrorists might attempt to attack or sabotage a nuclear power plant or other nuclear facility. Non-proliferation efforts, meanwhile, are concerned with preventing additional states acquiring nuclear weapons through the acquisition of the peaceful nuclear fuel cycle, including nuclear reactors, but more particularly enrichment and reprocessing facilities capable of producing material for nuclear devices. The case of Iran, which claims to want civilian nuclear power, but is clearly intent on acquiring at least the option of a nuclear weapon capability, in violation of its treaty obligations, is a contemporary example of the potential non-proliferation implications of a demand for nuclear energy by aspirant states. In this sense this book is also about the implications of a nuclear energy revival for international security more broadly defined.

For the purposes of this study 'global nuclear governance' refers to the web of international treaties, agreements, regulatory regimes, organizations and agencies, monitoring and verification mechanisms and supplementary arrangements that help determine the way that the peaceful uses of nuclear energy, notably the generation of nuclear electricity, is governed.[3] These may exist at the international, regional, sub-regional or bilateral levels. Governance at all these levels is in turn dependent on national implementation arrangements which ensure that each country fulfils its obligations in the nuclear field. Such a broad conceptualization of governance is intended to emphasize the need for a holistic approach when contemplating the implications of a civilian nuclear energy revival. Global governance will axiomatically be a collaborative enterprise involving many players. It will also be perpetually a work in progress.

Chapters 4 and 5 consider the current status and strengths and weaknesses of the existing global governance arrangements in the three key areas of nuclear safety, nuclear security and the non-proliferation of nuclear weapons. Chapter 6 analyses the likely impact of a nuclear energy revival on the three regimes. Although for the purposes of clarity this study treats nuclear safety, nuclear security and nuclear non-proliferation separately, it is increasingly recognized that there is an actual and potential relationship between them that is not always reflected in the ad hoc evolution of the global governance arrangements pertaining to each. Nor is it often reflected in policy or academic analysis. In particular the non-proliferation community on the one hand and the safety and security communities on the other tend to ignore each other. Seeking to overcome this intellectual 'stove-piping' was one of the motivations for including all of the regimes in this study.

While the uses of nuclear technology, including nuclear reactors, for peaceful purposes, are diverse, this study focuses on the generation of electricity by nuclear reactors. Although the global governance implications of the use of nuclear reactors for any type of peaceful purpose are the same – whether for desalination, process heat (such as for the production of oil from tar sands) or hydrogen production – the overwhelming majority of civilian nuclear reactors are used for generating electricity which is fed into power grids for domestic and industrial use. This study will therefore focus on this major use of nuclear reactors, mostly to the exclusion of others. This study thus does not consider developments pertaining to research reactors, isotope production reactors, experimental reactors or radioactive sources since these are not normally considered to be part of the nuclear revival, even though interest in them may be increasing. The study also does not consider nuclear reactors that are exclusively dedicated to producing plutonium for nuclear weapons, except in respect of the possibility that civilian nuclear power reactors might be converted to such purposes. Aspects of the nuclear fuel cycle beyond nuclear power plants themselves are not considered in detail but only where they are relevant to safety, security and non-proliferation.

The concluding chapter of this book provides ideas and recommendations for consideration by the international community as to how global governance might be strengthened in advance of a nuclear energy revival of whatever size. The role of the International Atomic Energy Agency (IAEA), the paramount organization in global nuclear governance, is highlighted. Particular attention is again paid to the period to 2030, although many of the recommendations would lay the groundwork for solid governance arrangements well beyond that. Finally, the likelihood that such recommendations for reform might actually be implemented is considered in the light of the considerable political, financial and technical barriers that they face. Among the most vexing of these is the stark and seemingly growing division between the nuclear 'haves' and 'have nots' over constraints on access to the peaceful uses of nuclear technology and the pace of nuclear disarmament.

1   Assessing a nuclear energy revival

The drivers

The three most important drivers of the current revival of interest in nuclear energy are: the perceived increasing global demand for energy, specifically electricity; the quest for energy security or diversity; and the need to tackle climate change. Technology, in the form of improved efficiency of existing reactors and the promise of advanced reactor design, is also an important driver. In addition there are political and strategic motivations that merit consideration.

Energy demand

One of the key drivers of a nuclear energy revival is said to be the projected overall increase in demand for energy, including electricity, worldwide. The assumption seems to be that nuclear will automatically maintain its current share in line with growth in energy demand overall, or even increase it. Several difficulties arise from such assumptions.

First, the growth in worldwide demand for energy is far from upwardly linear. While in 2008 the International Energy Agency (IEA) projected that world primary energy demand would increase by 45 per cent between 2006 and 2030, at an average annual rate of growth of 1.6 per cent (IEA 2008b: 78), in 2009 it reported that global energy use was set to fall 'for the first time since 1981 on any significant scale' (IEA 2009b: 42). Nuclear is unlikely to be immune to such perturbations in ever-upward energy projections.

Second, energy demand is regionally variegated rather than uniform around the globe, the current key drivers being in Asia and the Middle East (IEA 2009b: 42). National or regional energy demand projections cannot be necessarily extrapolated to demand for nuclear energy worldwide.

Third, basing projected demand for nuclear energy on overall energy demand is misleading since civilian nuclear power plants are devoted to providing electricity, which is only a part of overall energy consumption. Nuclear is currently absent from the vital transport sector, for instance.

Fourth, even basing nuclear power projections on global electricity demand is misleading. It seems undeniable that total global electricity demand will continue to rise over the long term due to global economic growth (once the current downturn ends), population increases and pressure from developing states for developed-world living standards. World electricity demand, despite the current downturn, is actually projected to grow at a higher rate annually than world primary energy demand at 2.5 per cent annually to 2030 (IEA 2009b: 42). Demand for electric cars to replace fossil fuel-powered vehicles may also boost demand (although there is debate about the extent of the likely increase) (*The Economist Technology Quarterly* 2009: 15).[1]

Yet global electricity demand estimates are often derived from simply adding up each country's projected demand for electricity, which in turn may be based in turn on predictions of national and economic growth, population growth and national greenhouse gas reduction targets. None of these indicators necessarily translates into increased demand for nuclear electricity. As with energy demand generally, electricity demand is also regionally and nationally variable. India's electricity demand is now estimated to be the highest in the world at 5.7 per cent annually to 2030 (IEA 2009b: 97), while Japan's is forecast to be only 0.7 per cent.

Finally, nuclear commonly provides baseload (non-peak) electricity, which is, again, only part of each country's overall installed electricity capacity. Since demand at peak periods is what drives the addition of new installed capacity (in order to prevent system overload), it may not be large nuclear power plants, which cannot be readily fired up or turned off, that are chosen for such additions.

Nuclear's global share of electricity production is in fact expected by the IEA to drop to 11 per cent in 2030 (IEA 2009b: 98). That share has gradually fallen from its historic peak of almost 18 per cent in 1996 to just below 14 per cent in 2008 (BP 2009). In terms of world electricity generating capacity, nuclear has also fallen from a peak of 12.6 per cent in 1990 to 8.4 per cent in 2007 (IAEA 2008e: 17, 47). This occurred not only because other forms of energy generation expanded faster, but because old nuclear power plants were being shut down and not replaced. Between 1990 and 2007, 73 new reactors were connected to the grid worldwide and 62 were closed, resulting in a net global increase of only 11 (NEA 2008: 47–49). Hence while there may be an increase in the total number of nuclear reactors worldwide in the coming decades, they are unlikely to generate an increasing percentage of electricity. Nuclear's competitors in baseload electricity production are likely to win out.

Energy security

A second driver of current increased interest in nuclear energy is the perceived need of states to ensure their energy security. If taken to mean complete national self-sufficiency in energy or 'energy independence' – such security is an illusion. No country in today's globalized world, with the possible exception of Russia, is able to be completely energy self-sufficient. Although governments and other observers often use the quest for energy security to make the case for nuclear power, what they are really calling for is more energy security or energy diversity. Diversity may in fact be the most important guarantee of energy security. As the Switkowski report on Australia's consideration of nuclear energy concluded, the most flexible and efficient national energy system 'is likely to include numerous technologies, each economically meeting the portion of the system load to which it is best suited ... A diversity of sources can also provide greater reliability and security of electricity supply' (Commonwealth of Australia 2006: 48).

Nuclear energy has some inherent drawbacks in helping achieve energy security, however defined. It cannot currently provide energy security to the vital transport sector. Nuclear power is also relatively inflexible in meeting peaks and troughs in electricity demand and cannot therefore replace other generation means like natural gas and coal if they suddenly become unavailable. Even France, which relies on nuclear for 77 per cent of its electricity, and which since the 'oil shocks' of the 1970s has had a deliberate strategy for achieving energy security, must fire up 40-year-old oil plants to meet peak electricity demand (Schneider 2009: 55). Mycle Schneider has calculated that France's real energy independence is 8.5 per cent of its total energy generation capacity, including nuclear (Schneider 2009: 64).

Uranium

The main argument in favour of nuclear energy providing energy security appears to be the current ready availability and cheapness of uranium, along with its energy intensity. This has traditionally been seen as one of the enduring advantages of nuclear energy. As the NEA puts it:

> The main advantages of nuclear power for energy security are the high energy density of uranium fuel combined with the diverse and stable geopolitical distribution of uranium resources and fuel fabrication facilities, as well as the ease with which strategic stockpiles of fuel can be maintained.
>
> (NEA 2008: 154)

These claims seem credible. One quarter of a gram of natural uranium in a standard fission reactor provides the same amount of energy as 16 kg of fossil fuel (MacKay 2009: 161). Stockpiling large strategic reserves of uranium is easier and cheaper than for oil, coal or gas, thereby avoiding the risk of a sudden shutdown of supply. While there are only a handful of major uranium suppliers, two of those with the largest reserves, Australia and Canada, are judged to be politically stable and commercially reliable (although others, like Kazakhstan and Niger, less so).

Moreover, uranium is ubiquitous and many countries have workable deposits. Global conventional reserves are estimated to be 4.7 million tons, with another 22 million tons in phosphate deposits and 4,500 million tons in seawater that are currently economically unrecoverable (MacKay 2009: 162). If the price of uranium ore rises above \$130 per kg, phosphate deposits would apparently become economic to mine (MacKay 2009: 162). Prices at that level would also stimulate exploration for traditional sources, as happened when the spot market price for uranium reached a record \$234 in December 2007 (it has since declined). According to the Nuclear Energy Agency (NEA) of the Organisation for Economic Cooperation and Development (OECD), sufficient uranium resources have been identified, if current usage rates continue, to provide 100 years of reactor supply (NEA 2008: 159). If the NEA's low scenario for expanded nuclear energy to 2030 of up to 404 GWe eventuates, the market will readily cope.

However, if the NEA's high scenario to 2030 of 619 GWe is accurate, the agency cautions that

> all existing and committed production centres, as well as a significant pro-portion of the planned and prospective production centres, must be com-pleted on schedule and production must be maintained at or near full capacity throughout the life of each facility.
>
> (NEA 2008: 164)

This seems a tall order in view of the recent record of uranium mining develop-ment delays, including in Australia and Canada, and may lead to uranium price rises in the future, followed inevitably by further bouts of exploration. Not all uranium deposits are easily mined, some are in such remote locations, like the middle of Western Australia, that they are currently too expensive to access, and others, as in the Democratic Republic of Congo, are located in areas of political instability.

Price, not just reliability of supply, is a key consideration in assumptions made about uranium as a source of energy security. Most studies concur that the cost of natural or enriched uranium will not be a barrier to increased use of nuclear energy. From 1983 to 2003 uranium was in a 20-year price slump and thus an energy bargain. This was partly due to over-production in the 1970s when the price was high, but also due to the availability of 'secondary' material held in various forms by civil industry and government, including military material (notably highly-enriched uranium (HEU)) and recyclable material (spent fuel). Secondary sources have supplied 40–45 per cent of the market in recent years (NEA 2008: 164). In addition the agreement between Russia and the United States for the down-blending of HEU from dismantled Russian nuclear weapons for use in American reactors will expire in 2013, making this 'secondary' source of uranium no longer available, unless the agreement is renewed. The surge in price between 2003 and 2007 was partly due to specula-tors in energy futures. Since most uranium sales are in the form of long-term contracts, the spot market volatility was misleading from an energy security per-spective. But the price rise also represented a 'market correction' based on the expectation of increased demand due to the anticipated nuclear energy revival.

Thorium

Thorium, about three times as abundant as uranium, is sometimes touted as a possible fuel for extending the future of nuclear power. Although not fissionable in its natural state, a fissionable form, thorium 233, can be created in a normal reactor if added to the uranium fuel, or can be bred in a fast breeder reactor. Experiments with a thorium fuel cycle have been conducted in several countries over the past 30 years, including Canada, Germany, India, Russia and the United States. India, which has four times as much thorium as uranium, is the most advanced in its plans. Norway, which has up to one-third of the world's thorium deposits, commissioned a report in 2007 by the Research Council of Norway, which concluded that 'the current knowledge of thorium based energy generation and the geology is not solid enough to provide a final assessment regarding the potential value for Norway of a thorium based system for long term energy production' (Research Council of Norway 2008). It recommended that the thorium option be kept open and that further research, development and international collaboration be pursued. According to the NEA: 'Much development work is still required before the thorium fuel cycle can be commercialized, and the effort required seems unlikely while (or where) abundant uranium is available' (WNA 2009e). It is clear that thorium will not be a commercially viable option in the coming decades.

Mixed oxide (MOX) fuel

Some in the nuclear industry suggest that the possibility of recycling plutonium reprocessed from spent nuclear fuel in the form of mixed oxide fuel (MOX) is one reason for a bright future for nuclear. MOX is composed of reprocessed plutonium and uranium (either natural or depleted). Thus a reprocessing capability is required. Only three countries currently have commercial-scale plutonium reprocessing plants, France (La Hague), Russia (Mayak) and the UK (Sellafield).

France is the pioneer and leading producer and user of MOX, deploying it in 20 light water reactors (LWRs) with up to 30 per cent of MOX fuel in their cores (Schneider 2009: 14). Altogether, some 39 conventional LWRs in Belgium, France, Germany and Switzerland operate with some MOX fuel (NEA 2008: 404). MOX has also been used to dispose of excess weapons-grade plutonium in commercial reactors in the United States. (For heavy water reactors like the Canadian CANDU that use natural uranium it is considered economically unattractive to reprocess spent fuel for MOX) (NEA 2008: 400).

Japan has had ambitious plans for using MOX for decades but has faced a series of difficulties in implementing them. It was only in October 2009 that MOX was loaded into a Japanese reactor for the first time (WNN 2009c). Japan has a small, underperforming pilot reprocessing plant at Tokai but most of its spent fuel has been reprocessed in France and the UK and the plutonium stored (both at home and abroad), making its stockpile of commercial plutonium the biggest in the world. Japan's full-scale Rokkasho reprocessing plant, long delayed and over budget ($20 billion or three times the estimated cost in 1993) (Smith 2007), is currently undergoing test operations. A MOX fuel fabrication plant is expected to be built by 2012. Japan plans to use plutonium for MOX in 16–18 existing reactors and later in its planned fast breeder reactors by 2015 (delayed from the original date of 2010) (Oshima 2009: 131).

The use of MOX has significant disadvantages for energy security. The fabrication process is potentially more hazardous than for uranium fuel, requiring expensive protective measures which increase the price (Garwin and Charpak 2001: 137). A 2003 MIT study concluded from its simple fuel cycle cost model under US conditions that 'the MOX option is roughly 4 times more expensive than once-through Uranium Oxide' (MIT 2003: 151). Spent fuel from MOX reactors is thermally hotter and more radiotoxic than spent uranium fuel (mainly in the form of minor actinides such as americium and curium[2]), as well as more voluminous (Paviet-Hartmann *et al.* 2009: 316), making it more difficult to dispose of in a repository (Bunn *et al.* 2003: 39). Unused plutonium from MOX spent fuel can be reprocessed and 'multi-recycled', but this 'becomes a burden on light water reactors because it yields less energy per kilogram of reprocessed fuel' (Garwin and Charpak 2001: 138). Only the fast breeder reactor can totally consume reprocessed plutonium and burn the minor actinides.

Reprocessing of reactor-grade plutonium, whether for MOX or use in fast reactors, has its own significant drawbacks. The Plutonium Uranium Extraction (PUREX) process, the most common method, generates massive volumes of waste. This has spurred efforts to develop new aqueous processes or radically different approaches, none of which are yet commercially proven (Paviet-Hartmann *et al.* 2009: 316).

Part of the argument for reprocessing is that it permits valuable energy sources in spent fuel, uranium and plutonium, to be re-used. Reprocessing is, however, much more expensive than the 'once-through' method and direct disposal of spent fuel, costing more than the new fuel is worth (von Hippel 2008). An official report commissioned by the French prime minister in 2000 concluded that using reprocessing instead of direct disposal of spent nuclear fuel for the entire French nuclear programme would be 85 per cent more expensive and increase average generation costs by about 5.5 per cent or $0.4 billion per installed GWe over a 40-year reactor life span (Charpin *et al.* 2000). For countries that have sent their spent fuel to France, the UK and Russia for reprocessing, the cost, at about $1 million per ton, is ten times that of dry storage (von Hippel 2008). The customer is, moreover, required by contract to take back the separated plutonium and other radioactive waste. The three commercial reprocessing countries have thus lost virtually all of their foreign customers, making their reprocessing plants more uneconomic than they were before. The UK proposes to shut Sellafield in the next few years at a cost of $92 billion, including cleanup of the site (von Hippel 2008).

Even the argument that the amount of waste is reduced through reprocessing is suspect. In fact 'recycling' plutonium 'only reduces the high-level waste problem minimally' (von Hippel 2008). In France used MOX fuel still contains 70 per cent of the plutonium it did when manufactured, so it is returned to the reprocessing plant for further reprocessing. Thus France is, in effect, using reprocessing to move its spent fuel problem from reactor site to reprocessing plant and back again (von Hippel 2008).

Finally, MOX also carries proliferation risks since, despite earlier assumptions to the contrary, even non-weapons grade material can be used in a crude nuclear weapon. A 1994 report by the Committee on International Security and Arms Control (CISAC) of the US National Academy of Sciences, called *Management and Disposition of Excess Weapons Plutonium*, showed that it is much easier to extract plutonium from fresh MOX fuel than from spent fuel. Hence MOX fuel must be closely monitored to prevent diversion (Committee on International Security and Arms Control 1994). The complications associated with this technology constitute a constraint on expansive plans for nuclear energy beyond the once-through system in which spent fuel is simply disposed of.

Nuclear technology dependence

Perhaps the most telling counterpoint to the alleged energy security that nuclear energy provides is the fact that, apart from uranium, the entire civilian nuclear fuel cycle is supplied by a small number of companies and countries. Nuclear reactor technology, nuclear reactor manufacturing techniques and skills, fuel fabrication, uranium enrichment and reprocessing are concentrated in few hands, making most countries more, rather than less, dependent on others for this energy source. The nuclear power plant construction industry has, moreover, seen significant consolidation and retrenchment over the last 20 years. Even the United States no longer has a 'national' nuclear reactor manufacturing capability after the nuclear divisions of Westinghouse were sold to British Nuclear Fuels in 1999 and then to Toshiba of Japan in 2006 (NEA 2008).

Various takeovers and mergers have resulted in just two large consolidated nuclear power plant vendors, Westinghouse/Toshiba and Areva NP of France. Even Areva only makes the reactors themselves and must partner with others, like Siemens or Electricité de France (EDF), to build entire nuclear power plants. Mitsubishi Heavy Industries, the new vertically integrated Russian company Atomenergoprom (including Rosatom) and Canada's Atomic Energy of Canada Limited (AECL) are lesser players. As the NEA points out, 'There is no single company that can build a complete nuclear power plant by itself' (NEA 2008), although new players may emerge that are capable of doing so.

Some individual countries can and do build and export nuclear reactors by themselves, despite the multinational nature of the nuclear industry. A South Korean conglomerate in 2010 that includes the Korea Electric Power Company (KEPCO) secured its first contract to build nuclear power plants abroad, in the United Arab Emirates (UAE). Russia, China and, farther in the future, India, are mooted as potential new major national reactor suppliers.

This situation renders most states mere importers of materials, skills and technology and therefore subject to the decisions of exporters and their governments, whether on political, commercial or non-proliferation grounds. On non-proliferation grounds alone there are significant constraints on states acquiring the full nuclear fuel cycle, whether uranium enrichment at the 'front end' or spent fuel reprocessing at the 'back end'. The Nuclear Suppliers Group (NSG), a group of nuclear technology and materials exporting countries, is currently seeking

Table 1.1 The international nuclear industry

	Areva	Westinghouse-Toshiba	General Electric-Hitachi	Rosatom	AECL	KEPCO
Headquarters	France	United States	United States/Japan	Russia	Canada	South Korea
Ownership structure	87% French government; 13% private sector	67% Toshiba; 20% Shaw Group; 10% Kazatomprom	Hitachi owns 40% of GE and GE owns 20% of Hitachi	100% Russian government	100% Canadian government	21% Korean government; 30% Korea Finance Corporation; 25% Foreign; 24% Other
Reactor type	Pressurized Light Water EPR-1000	Pressurized Light Water AP-1000	Boiling Water ABWR	Pressurized Light Water VVER-1200	Pressurized Heavy Water CANDU	Advanced Pressurized Reactor 1400
Reactors operating	71	119	70	68	30	0
Reactors under construction	6	5	4	16	0	2

Source: reproduced (with permission) and adapted from 'The Global Nuclear Revival' in Duane Bratt, *Canada and the Global Nuclear Revival* (forthcoming).

agreement among its members to further constrain the export of sensitive enrichment technologies to additional countries. There are also continuing attempts to establish multilateral mechanisms, such as an IAEA 'fuel bank', to assure states with nuclear power that they will always be able to obtain the necessary fuel without resorting to enrichment themselves. The vast majority of states with nuclear reactors will therefore continue to be dependent on importing both fuel and nuclear technology. This should be sufficient to convince policy-makers that while nuclear power can add to national energy diversity, and may provide additional energy security in the sense of relative security of fuel supply, it cannot provide the elusive energy independence.

Climate change

One of the arguments increasingly used to promote nuclear power is the need to tackle climate change. The British Labour Party government, in laying out the case for 'new build' in the UK, used this justification most explicitly of any government. Business and Enterprise Minister John Hutton asserted in 2008 that 'Set against the challenges of climate change and security of supply, the evidence in support of new nuclear power stations is compelling' (WNN 2008f). Some 'Greens', notably the former founding member of Greenpeace, Patrick Moore (Moore 2006: BO1), and British atmospheric scientist James Lovelock, father of the Gaia theory, have been converted to a pro-nuclear stance on the grounds that climate change is so potentially catastrophic that all means to reduce greenhouse gases must be utilized (Norris 2000). Pro-nuclear energy non-governmental organizations have emerged to campaign for increased use of nuclear energy, such as Environmentalists for Nuclear Energy and the US-based Clean and Safe Energy Coalition.

Nuclear power, like hydropower and other renewable energy sources, produces virtually no carbon (CO_2) directly. The NEA's *Nuclear Energy Outlook* notes of nuclear that 'On a life-cycle basis an extremely small amount of CO_2 is produced indirectly from fossil fuel sources used in processes such as uranium mining, construction and transport' (NEA 2008: 121). The generation of nuclear electricity does, however, in addition, emit carbon by using electricity from the grid for fuel fabrication, the operation of nuclear power plants themselves, and in other aspects of the nuclear fuel cycle, especially enrichment and reprocessing. It is not therefore as entirely carbon-free as some observers claim.

To date the international climate change regime has not been particularly favourable to nuclear energy. Under the 1997 Kyoto Protocol to the 1992 United Nations Framework Convention on Climate Change (UNFCCC) states may use nuclear power to help them meet their own greenhouse emission targets, but the regime has not permitted them to build nuclear power plants in developing countries in order to obtain certified emissions credits under the so-called clean development mechanism (CDM). This was due to strong opposition to nuclear energy from influential state parties on the grounds of sustainability, safety, waste disposal and weapons proliferation.

Although the December 2009 Copenhagen climate summit failed to agree on a new regime, one is eventually likely to emerge that includes deeper mandated cuts, the involvement of a broader range of states in such cuts and, potentially, a global carbon cap-and-trade system (accompanied in some states by a carbon tax). The latter would be favourable to nuclear energy. Nuclear energy may even find greater official encouragement in a new climate change treaty, due to the growing urgency of preventing climate change and the argument that it needs tackling with any means possible. Changes in the attitude of some influential governments about nuclear power, like Italy, Sweden and the UK, may help propel this.

The Intergovernmental Panel on Climate Change (IPCC) has meanwhile reached the startling conclusion that to stabilize global temperatures at two degrees above pre-industrial levels (widely regarded as the only way to avoid potentially catastrophic consequences) would require greenhouse emissions to be cut by 50–85 per cent below 2000 levels by 2050 (IPCC 2007). Scenarios devised by international agencies for doing this all propose a significant role for nuclear on the grounds that it is one of the few established energy technologies with a low-carbon footprint.

A renowned study by Pacala and Socolow published in the scientific journal *Science* in 2004 demonstrated how current technologies, including nuclear energy, could help reduce carbon emissions by seven billion tons of carbon per year by 2050 through seven 'wedges' of one billion tons each (Pacala and Socolow 2004). The nuclear wedge, 14.5 per cent of the total, would require adding 700 GWe capacity to current capabilities, essentially doubling it, by building about 14 new plants per year. While this is a reasonable rate (the historical annual high was around 33 reactors in 1985 and 1986 (NEA 2008: 316)), the Pacala/Socolow estimates did not take into account that virtually all existing reactors will have to be retired by 2050, even if their operating lives are extended to a total of 60 years (Squassoni 2009a: 25). Thus 25 new reactors in total would have to be built each year through 2050 to account for retirements.

The International Energy Agency (IEA), another part of the OECD, in its 2008 *Energy Technology Perspectives*, suggested that as part of its radical Blue Map scenario there should be a 'substantial shift' to nuclear to permit it to contribute 6 per cent of CO_2 savings, considerably lower than the 14.5 per cent wedge, based on the construction of 24–43 1,000 MW nuclear power plants each year (32 GWe of capacity) between now and 2050 (IEA 2008a: 41). The figures differ from the Pacala/Socolow wedge analysis because the Blue Map envisages higher carbon levels by 2050 and more severe cuts in carbon (half of 2005 levels rather than a mere return to such levels). The IEA implied that not all countries would need to choose nuclear, noting that 'considerable flexibility exists for individual countries to choose which precise mix of carbon capture and storage (CCS), renewables and nuclear technology they will use' (IEA 2008a: 42). It called, however, for nothing less than an energy revolution, arguing that 'Without clear signals or binding policies from governments on CO_2 prices and standards, the market on its own will not be sufficient to stimulate industry to act with the speed or depth of commitment that is necessary' (IEA 2008a: 127).

IEA recommendations for achieving greenhouse gas (GHG) targets by 2050 are relevant as a driver of interest in nuclear energy, but industry would need to gear up now to sustain the substantial and steady increase envisaged and it would still have to compete with alternative technologies for achieving carbon abatement. The NEA posits two scenarios to 2030 (IEA 2008a: 134–135). Its low estimate projects that nuclear will displace only slightly more carbon per year than it does now, estimated at 2.2–2.6 Gt of coal-generated carbon (IEA 2008a: 123). This assumes that carbon capture and storage (CCS) and renewable technologies are successful, 'experience with new nuclear technology is disappointing', and that there is 'continuing public opposition to nuclear power'. The high scenario for 2030, on the other hand, projecting almost 5 Gt of carbon displacement, assumes that 'experience with new nuclear technologies is positive', and 'there is a high degree of public acceptance of nuclear power'. A 2003 study by the Massachusetts Institute of Technology (MIT) estimated that a three-fold expansion of nuclear generating capacity to 1,000 gigawatts (GW) by 2050 would avoid about 25 per cent of the increment in carbon emissions otherwise expected in a business-as-usual scenario (MIT 2003: ix).

These hedged scenarios reveal that the barriers to nuclear contributing significantly to meeting GHG reduction targets are two-fold: technological and political. Opinions differ as to how high these barriers are. Members of the 2007 Keystone Nuclear Power Joint Fact-Finding Dialogue – drawn from a broad range of 'stakeholders', including the utility and power industry, environmental and consumer advocates, non-governmental organizations, regulators, public policy analysts and academics – reached no consensus on the likely rate of expansion of nuclear power over the next 50 years in filling a substantial portion of its assigned carbon 'wedge' (Keystone Center 2007: 10). The MIT study recommended 'changes in government policy and industrial practices needed in the relatively near term to retain an option for such an outcome' (MIT 2003: ix), but in a 2009 review of its earlier report despaired at the lack of progress (MIT 2009: 4).

On the political side, there appears to be consensus that a 'business-as-usual' approach to nuclear energy will not increase its contribution to tackling climate change. Nuclear's long lead times (reactors take up to ten years to plan and build) and large upfront costs, compared to other energy sources and energy conservation measures, mean that without a determined effort by governments to promote nuclear it would have little impact by 2030 in reducing GHG emissions. Even replacing the existing nuclear fleet to maintain the current contribution of nuclear to GHG avoidance will require a major industrial undertaking in existing nuclear energy states. Despite the rhetoric, there is scant evidence that governments are taking climate change seriously enough to effect the 'energy revolution' that the IEA has called for, much less taking the policy measures that would promote nuclear energy as a growing part of the solution.

Even if carbon taxes or emissions trading schemes help level the economic playing field by penalizing electricity producers that emit more carbon, these measures are likely to take years to establish and achieve results. They will also benefit, probably disproportionately, cheaper and more flexible low- or non-carbon emitting technologies such as renewables, solar and wind, and make conservation and efficiency measures more attractive (see Chapter 2 for further analysis of the effects of carbon taxes and/or emissions trading schemes on the economics of nuclear power).

One of the seemingly plausible arguments in favour of using nuclear to tackle climate change is that the problem is so potentially catastrophic that every means possible should be used to deal with it – by implication regardless of cost. However, this ignores the fact that resources for tackling climate change are not unlimited. Already governments and publics are baulking at the estimated costs involved. Therefore, the question becomes what are the most economical means for reducing a given amount of carbon. One way of answering this is to examine the per dollar cost of reducing coal-fired carbon emissions through various alternative means of generating electricity.

According to research by Lovins and Sheikh, nuclear power delivers less 'electrical service' per dollar than its rivals, as Figure 1.1 illustrates. In their estimation nuclear surpasses only centralized, traditional gas-fired power plants burning natural gas at relatively high prices in terms of carbon emissions displaced per dollar spent. Large wind farms and cogeneration are 1.5 times more cost-effective than nuclear in displacing carbon. Efficiency measures are reportedly about ten times more cost-effective in reducing carbon than nuclear. In sum

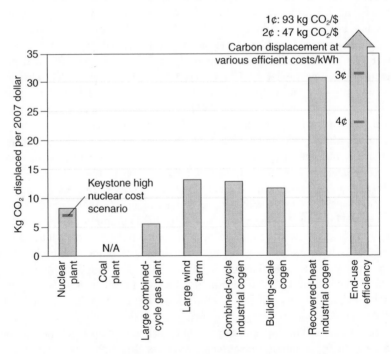

Figure 1.1 Coal-fired CO_2 emissions displaced per dollar spent on electrical services (source: Lovins and Sheikh 2008, reproduced with permission).

they claim that 'every dollar spent on nuclear power will produce 1.4–11+ times less climate solution than spending the same dollar on its cheaper competitors' (Lovins and Sheikh 2008: 16). While such studies are naturally dependent on the variables used and assumptions made, this study has not been able to find a competing chart, official or otherwise, that calls into question such conclusions.

This is where the opportunity cost argument has sway: put simply, it is not possible to spend the same dollar on two different means of tackling climate change at once. Although such considerations have not yet seeped into political consciousness in many countries, this is increasingly likely to happen as the price of alternatives drops and governments focus on 'big wins', the measures that will have the greatest impact at the lowest price. What happens after the 'low hanging fruit' are picked is a different question, but no country is anywhere near that point yet.

There is a possibility that runaway global warning will become more apparent and politically salient through a catastrophic event like a sudden halt to the North Atlantic sea current or the disappearance of all summer ice from the North Pole. Since the 2008 IPCC report was released, a growing number of climatologists has concluded that the report underestimated both the scale and pace of global warming, notably changes in the Arctic ice sheet and sea levels. NASA's Jim Hansen, 'perhaps the world's foremost climatologist', has calculated that the situation is so dire that 'the entire world needs to be out of the business of burning coal by 2030 and the Western world much sooner' (Hansen *et al.* 2008: 217–231).[3] In such circumstances massive industrial mobilization to rapidly build nuclear power plants, along the lines of the Manhattan Project to build the US nuclear bomb or the Marshall Plan for European economic recovery after the Second World War, may be politically and technologically desirable.

But nuclear power, as indicated in the discussion of constraints below, would still face numerous barriers in responding to such a catastrophe that other alternatives do not. As Sharon Squassoni notes, 'If major reductions in carbon emissions need to be made by 2015 or 2020, a large-scale expansion of nuclear energy is not a viable option' for that purpose (Squassoni 2009a: 28). It is simply too slow and too inflexible compared to the alternatives. As the 2007 Keystone report noted, just to build enough nuclear capacity to achieve the carbon reductions of a Pacala/Socolow wedge 'would require the industry to return immediately to the most rapid period of growth experienced in the past (1981–1999) and to sustain this rate of growth for 50 years' (Keystone Center 2007: 11, 27).

There is a further climate change twist for nuclear: the vast amounts of water that nuclear reactors normally need for cooling purposes. If climate change reduces river flow or results in warmer water, new nuclear power plants will have to be located on sea coasts. Plant costs can reportedly change by $1 billion depending on whether a plant is cooled by saltwater or freshwater (MacLaclan 2009a). Plants already using river water may be forced to close or require costly changes to avoid over-heating water that is to be discharged back into increasingly warm rivers. France has already been forced to shut down certain reactors during heat waves for this reason. The Indian Point reactor in upstate New York is currently facing closure unless it undergoes expensive modifications to avoid its discharge killing thousands of fish every year in the Hudson River (*New York Times* 2010). Technology that uses less cooling water is being developed for new reactors, but retrofits to old ones may be necessary (and costly) as well. This is not to say that other power generation technologies will be unaffected by climate change (hydroelectricity and gas recovery from shale in particular require plentiful water supplies), but nuclear is not itself immune.

The promise of nuclear technology – current and future

While the notion of a mechanistic 'technological imperative' is now discredited, in the current case of renewed interest in nuclear energy the promise of improved technology is at least a partial driver. The following section considers both the technologies themselves and programmes designed to research and promote new technologies.

Improvements in current power reactors (Generation II)

Most reactors in operation today are 'second generation', the first generation having been largely experimental and unsuited for producing significant grid electricity. The global fleet is dominated, both in numbers and generating capacity, by light water reactors derived from US technology originally developed for naval submarines. These comprise two types: pressurized water reactors (PWR) and boiling water reactors (BWR). The rest are pressurized heavy water reactors (PHWR) based on the Canadian Deuterium Uranium (CANDU) type; gas-cooled reactors (CCRs); or Soviet-designed light-water-cooled, graphite-moderated reactors (LWGRs or RBMKs – the Russian acronym for high power channel reactor). Most current reactors operate on the 'once-through' system: the original fuel, low enriched uranium (LEU) or natural uranium, is used in the reactor once and the spent fuel treated as waste and stored rather than being reprocessed for further use.

Significant improvements have been made in the past few decades in existing reactor operations, leading to higher fuel burn-up and improved capacity factors. According to the NEA, these developments have resulted in savings of more than 25 per cent of natural uranium per unit of electricity produced compared to 30 years ago, and a significant reduction in fuel cycle costs (NEA 2008: 401). In addition to higher 'burn-up' rates, current nuclear reactors worldwide are also being overhauled and receiving significant life extensions rather than being closed down. Given that much of the initial capital investment has been paid or written off, such reactors are highly profitable. This improvement in the profitability of existing reactors has been one of the drivers of renewed interest in nuclear energy, although the assumption that new reactors would be equally profitable is yet unproven.

Generation III and Generation III+ reactors

One of the main technological arguments for a nuclear revival rests on the emergence of so-called Generation III and Generation III+ reactors. According to their manufacturers, these types promise several advantages over Generation II: lower costs through more efficient fuel consumption and heat utilization; a bigger range of sizes; and increased operational lifetimes to approximately 60 years. They are also reportedly able to operate more flexibly in response to customer demand. Perhaps most important, they are reputedly safer, incorporating 'passive' safety systems that rely on natural phenomena – such as gravity, response to temperature or pressure changes and convection – to slow down or terminate the nuclear chain reaction during an emergency. This contrasts with the original designs, which relied on human intervention.

The industry is also promising that economies of scale through standardization, modular production techniques and advanced management systems will bring prices down after the initial first-of-a-kind (FOAK) plants have been built. Nuclear Energy Institute (NEI) President and CEO Marvin Fertel claims that 'If you are using standardized plants, everything from licensing to construction isn't a ten-year period anymore', supposedly resulting in a much greater rate of deployment in the decade 2020 to 2030 than in the decade 2010 to 2020 (Weil 2009b: 4). This implies that the real revival is likely to emerge at least a decade from now. Economies of scale are also premised on the size of the reactors: since construction cost is the biggest factor in the price of a nuclear power plant, building a bigger one that produces more electricity is said to reduce the 'levelized' cost of the power produced. These issues are discussed extensively in the section on the economics of nuclear power in Chapter 2.

The distinction between Generation III and Generation III+ seems arbitrary and more a question of marketing strategy than science. According to the US Department of Energy (DOE), Generation III+ reactors promise 'advances in safety and economics' over Generation III (DOE 2009).[4] The NEA suggests vaguely that Generation III+ designs are 'generally extensions of the Generation III concept that include additional improvements and passive safety features' (NEA 2008: 373–374). It advises that:

> the difference between the two should be defined as the point where improvements to the design mean that the potential for significant off-site releases [of radioactivity] has been reduced to the level where an off-site emergency plan is no longer required.

Several companies in France, Japan, South Korea and the United States are developing self-described Generation III or Generation III+ designs for light water reactors. Other countries, notably Canada, China and Russia, have plans to produce their own type of Generation III or Generation III+ reactors. China plans to 'assimilate' the AP1000 technology and 're-innovate' its design, but

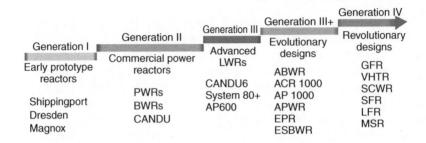

Figure 1.2 Progression of nuclear reactor technologies (source: adapted from NEA 2008: 373).

in addition has its own second-generation CPR-1000 reactor, derived from French designs imported in the 1980s. It hopes to build both designs en masse in China (WNN 2009b). Canada is developing an Advanced CANDU Reactor (ACR), the ACR1000, based on its original pressurized heavy water reactor, but using slightly enriched rather than natural uranium. Since 2008 South Korea has been able to design, manufacture and build its own reactors, one of which is the APR1400 that is being built domestically and is being sold to the UAE.

Although all of these designs are 'evolutionary' rather than 'revolutionary', their performance is to date unproven since they are, with the exception of four Advanced Boiling Water Reactors (ABWRs) in Japan, not yet in existence.

Small, medium-sized, miniaturized and other novel reactors

Much has been made of the need and likely market for small- and medium-sized reactors (SMRs), below the current trend of 1,000 MW and above. The IAEA officially defines small reactors as those with a power output of less than 300 MWe, while a medium reactor is in the range of 300–700 MWe.[5] Although such reactors have been investigated, developed and in some cases deployed since the 1950s, there is currently renewed interest.

The NEA reports that some 60 different types of SMRs are 'being considered' globally, although none has yet been commercially established (NEA 2008: 380). Countries involved in researching them include Argentina, Canada, China, India, Japan, Russia, South Korea and the United States (IAEA 2009r). Currently only India has successfully utilized such types of units with its domestically produced 200 and 480 MWe heavy water reactors. SMRs are currently

Table 1.2 Small- and medium-size reactors

In operation	133
Under construction	12
Number of countries with SMRs	28
Generating capacity, GW(e)	60.3

Source: IAEA 2010c.

being constructed by Russia in India and Romania. Several other states are reported by the IAEA to be ready to deploy SMR designs including Canada and China (IAEA 2009r).

SMRs are advertised as overcoming all of the current barriers to wider use of nuclear energy, especially by developing countries. They are said to be ideal for countries with relatively small and undeveloped electricity grids, as well as huge countries like Russia with large areas unconnected to the national grid. They promise to be cheaper and quicker to build, installable in small increments as demand grows, and able to be sited close to population centres, thereby reducing the need for long transmission lines. They can allegedly be more safely connected to smaller grids, operate off-grid, or be used directly for heating, desalination or hydrogen production. Some SMR designs reportedly would have 'reduced specific power levels' that allow for design simplification, thereby enhancing safety and reliability and making them 'especially advantageous in countries with limited nuclear experience' (NEA 2008: 380). Half of the SMR designs under consideration are planned without on-site refuelling, in some cases due to the use of higher enriched fuel than a normal light water reactor (NEA 2008: 380). SMRs are seen as 'an elegant solution to problems requiring autonomous power sources not requiring fuel delivery in remote locations' (Statens Strålevern 2008: 6).

Such reactors remain, however, mostly in the research and development phase (NEA 2008)[6] and will remain unproven for many years in respect of their reliability, safety, security and weapons non-proliferation potential. One of the alleged leading designs, the South African pebble bed reactor, has turned out to have significant safety problems that will likely delay its deployment by years (Thomas *et al.* 2007). China has reportedly re-engineered the pebble bed design to make it safer and more reliable and is moving to the prototype stage before deploying it domestically and potentially exporting it (WNA 2010a).

The economic and commercial viability of SMRs remains unproven. Although they might be cheaper per unit than a large reactor, they miss out on economies of scale that the nuclear industry touts as a major benefit of 1,000 MW+ reactors and may thus produce electricity at a higher levelized cost. They would certainly be more suited to modern 'distributive' electricity systems, where generating plants are located closer to the customer, but they may face local political opposition from publics which are content to rely on nuclear power as long as it is kept distant, but who might baulk at even a small plant 'in their backyard'. While the NEA predicts that some SMRs will be available commercially between 2010 and 2030, such a vague 20-year timeframe does not inspire confidence.

The practicalities and economics of small reactors may be more favourable for off-grid applications such as heating and providing electricity in small communities or for incremental grid additions, especially if they are to be built assembly-line style (Hiruo 2009). This is apparently the motivation behind Russia's construction of small floating reactors that will be towed to deployment sites (Statens Strålevern 2008: 55), although again the economics and safety and security of such portable units has not yet been established (Kramer 2010). In sum, while there appears to be great interest in and R&D activity focused on SMRs, the prospects of commercialization remain unclear and therefore their contribution in the coming decades to meeting global electricity demand equally uncertain.

Generation IV reactors

Generation IV reactors promise revolutionary advances on Generation III+ models. But as Ian Davis puts it, Generation IV reactors are 'revolutionary' only in the sense that they 'rely on fuel and plants that have not yet been tested' (Davis 2009: 19). Generation IV reactors in most cases will seek to 'close' the nuclear fuel cycle, leading to higher energy usage per amount of uranium or recycled fuel, less nuclear waste due to the more efficient burning of plutonium and other highly radioactive actinides, reduced capital costs, enhanced nuclear safety and less weapons proliferation risk. It is envisaged that such reactors will rely on new materials and metals yet to be developed, including those able to resist corrosion far in excess of today's levels.

Research and development for such designs is expensive, which has led to international cooperative efforts to share the burden. Nine countries and Euratom formed the Generation IV International Forum (GIF) in 2001, under the auspices of the NEA, to develop new systems 'intended to be responsive to the needs of a broad range of countries and users' (GIF 2008). Now with 12 country members plus Euratom,[7] GIF has chosen the six most promising systems below to investigate further, although none are expected to be deployable until at least 2025.

The International Project on Innovative Nuclear Reactors and Fuel Cycles (INPRO), established under IAEA auspices in 2001, also aims to help by bringing together technology holders and users to 'consider jointly the international and national actions required for achieving desired innovations in nuclear reactors and fuel cycles' (IAEA 2009g). As of August 2009 30 countries and the European Commission had joined (IAEA 2009g).[8]

None of the Generation IV reactors are expected to be available in the coming couple of decades (Sub-Committee on Energy and Environmental Security 2009: 13). As the World Business Council for Sustainable Development puts it, such technology is 'promising but far from being mature and competitive' (World Business Council for Sustainable Development 2008: 16).

- Gas-Cooled Fast Reactor (GFR): features a fast-neutron-spectrum, helium-cooled reactor and closed fuel cycle;
- Very-High-Temperature Reactor (VHTR): a graphite-moderated, helium-cooled reactor with a once-through uranium fuel cycle;
- Supercritical-Water-Cooled Reactor (SCWR): a high-temperature, high-pressure water-cooled reactor that operates above the thermodynamic critical point of water;
- Sodium-Cooled Fast Reactor (SFR): features a fast-spectrum, sodium-cooled reactor and closed fuel cycle for efficient management of actinides and conversion of fertile uranium;
- Lead-Cooled Fast Reactor (LFR): features a fast-spectrum lead of lead/bismuth eutectic liquid-metal-cooled reactor and a closed fuel cycle for efficient conversion of fertile uranium and management of actinides;
- Molten Salt Reactor (MSR): produces fission power in a circulating molten salt fuel mixture with an epithermal-spectrum reactor and a full actinide recycle fuel cycle.

(Generation IV International Forum, www.gen-4.org/, accessed 2 June 2010)

Fast neutron and breeder reactors

A fast neutron reactor differs from a traditional thermal reactor in using for its core a composite of natural uranium, uranium-238, with about 10 per cent plutonium or enriched uranium. Such a reactor can produce up to 60 times more energy from uranium than thermal reactors (NEA 2008: 80). Fast reactors can operate in two ways. If the amount of HEU is limited, they operate in 'burner' mode and are able to dispose of redundant nuclear and radioactive material. Some nuclear scientists have advocated recycling spent fuel in fast reactors as a way of dealing with the nuclear waste problem while also improving utilization of the energy source (Hannum *et al.* 2005). Since a fast reactor has no moderator, it is compact compared to a normal reactor. A 250 MWe prototype in the UK had a core 'the size of a large dustbin' (NEA 2008: 450). Such reactors are usually cooled by liquid sodium, which is efficient at removing heat and does not need to be pressurized.

With the addition of an extra uranium 'blanket', normally depleted uranium, fast neutron reactors have the remarkable quality of 'breeding' more plutonium than they use. Operated in this way they are known as fast breeder reactors (FBR) and are the basis for the notion that states could acquire the ultimate in energy security and operate a 'plutonium economy' based on an endless supply of fuel. According to the NEA, FBRs could thus potentially increase the available world nuclear fuel resources 60-fold (NEA 2008: 450).

There is currently only one operational FBR, in Russia, although around 20 have been built and operated in a handful of countries at various times (NEA 2008: 450). France's Phénix reactor was permanently shut down as recently as October 2009. China, India and Japan are attempting to develop FBRs. Japan plans to operate a demonstration reactor by 2025 and a commercial model by 2050 (Oshima 2009: 131). In the meantime, its shutdown Monju reactor was restarted in May 2010 after design changes and a safety review following an accident and cover-up 14 years ago (Yamaguchi 2010). But as Kenichi Oshima notes, 'In reality, however, the fast breeder reactor development has been delayed repeatedly and there is no chance of it being feasible' (Oshima 2009: 131).

The history of fast neutron and breeder reactors is discouraging. In the 1960s and 1970s, the leading industrialized countries, including the United States, put the equivalent of more than $50 billion into efforts to commercialize fast neutron reactors in the expectation that they would quickly replace conventional reactors (von Hippel 2008). They have proved costly, unreliable and accident-prone due to the explosive nature of sodium on contact with air or water. Serious accidents occurred at the Fermi reactor near Detroit in 1966 and at the Monju reactor in 1995. The French Superphénix breeder reactor closed in 1998 with an effective lifetime capacity factor of just 6.3 per cent (Davis 2001: 290) after 12 years of operation during which it intermittently delivered up to 1,200 MWe to the grid. It is the only breeder reactor that has ever been capable of producing electricity comparable to the largest thermal reactors commonly in operation. According to Garwin and Charpak, 'The decision to build an expensive industrial prototype breeder in France was premature; it was due in part to technological optimism on the part of the participants, coupled with a lack of appreciation for alternatives' (Garwin and Charpak 2001: 135).

Advanced fast breeders are among the Generation IV designs, including those being considered under GIF. Such reactors will not, however, be technologically or commercially viable on a large enough scale to make a difference to the provision of nuclear energy to electricity grids in the coming decades. Dreams of a self-sustaining 'plutonium economy', the ultimate in energy security, in which breeder reactors provide perpetual fuel without the need for additional imports of uranium, are likely to remain dreams.

Fusion power

Barring a technological miracle, fusion reactors will also not be available for the foreseeable future and will contribute nothing to the current nuclear revival (US Presidential Committee of Advisors on Science and Technology 1995). Research and development, at enormous cost, is nonetheless continuing. The International Thermonuclear Experimental Reactor (ITER), financed by six countries, China, India, Japan, South Korea, Russia and the United States, and the EU, is being built in Cadarache, France. The US National Ignition Facility (NIF) in Livermore, California, began operations in May 2009 at a cost so far of $4 billion,

Table 1.3 Fast breeder reactors by status and country, 2009

Unit	Country	Status	Construction date	Shutdown date	Power output (MWe)
Beloyarsky-3	Russian Federation	Operational	1969		560
PFBR	India	Under construction	2004		470
Beloyarsky-4	Russian Federation	Under construction	2006		750
Phénix	France	Shut down	1968	2009	130
Super-Phénix	France	Shut down	1976	1998	1,200
Kalkar (KKW)	Germany	Cancelled	1973	1991	295
KNK II	Germany	Shut down	1974	1991	17
Monju	Japan	Online	1986	1995	246
BN-350	Kazakhstan	Shut down	1964	1999	52
South Urals 1	Russian Federation	Suspended	1986	1993	750
South Urals 2	Russian Federation	Suspended	1986	1993	750
Dounreay DFR	United Kingdom	Shut down	1955	1977	11
Dounreay PFR	United Kingdom	Shut down	1966	1994	250
Enrico Fermi-1	United States	Shut down	1956	1972	60

Sources: Power Reactor Information System (PRIS), International Atomic Energy Agency, 2009 (accessed 3 May 2010). Available at: www.iaea.org/programmes/a2/ (Yamaguchi 2010).

almost four times the original estimate, and is more than five years behind schedule. Its principal role is to simulate thermonuclear explosions as part of the stockpile stewardship programme for US nuclear weapons, but it can also be used for fusion energy experiments. In the case of both programmes, full experiments to test nuclear fusion as a power source seem likely to be delayed until 2025 (*The Economist* 2009c: 82). The IEA contends that 'Fusion is not likely to be deployed for commercial electricity production until at least the second half of the century' (IEA 2008a: 306).

Political and promotional drivers

Wider considerations than those analysed above will always come into play for government policy-makers when considering nuclear energy. A decision to launch or significantly expand a nuclear power programme is hugely complex, involving an array of international and domestic political legal, economic, financial, technical, industrial and social considerations. Nuclear energy has been so controversial in the past and involves so many stakeholders that it is the quintessential candidate for politicization, to the chagrin of those seeking 'rational' energy policies.

Political drivers

The political drivers that may motivate a state to seek nuclear energy for the first time include: the quest for national prestige; a perceived need to demonstrate a country's prowess in all fields of science and technology; a predisposition towards high-profile, large-scale projects of the type that nuclear represents; and the desire for modern, cutting-edge technology no matter how suited or ill-suited to an individual country's requirements. These considerations do not generally apply to the acquisition of other forms of electricity generation and may make the difference in decision-making. Current intimations by such states as Ghana, Nigeria and Venezuela that they are considering nuclear power probably represent a triumph of national ambition over sound energy policies. As Lowbeer-Lewis notes of Nigeria's plans:

> Nigeria views itself as a regional power. The successful development of a nuclear energy program can thus be partially perceived as a matter of national pride and a means of cementing the country's status as a leader on the African continent.
>
> (Lowbeer-Lewis 2010: 3)

But in considering the likelihood of a nuclear revival the possibility that such states may actually succeed cannot be discounted. Autocratic states, partially democratic states and/or those with command economies may be able to politically override other considerations weighing against nuclear energy. Venezuelan President Hugo Chavez, for example, in nurturing nuclear cooperation arrangements with Iran, France and Russia, may actually lead his oil-rich country towards acquiring an energy source that, on the face of it, it does not need.

'Nuclear hedging', based on a belief that a nuclear power programme will help provide the basis for a future nuclear weapon option, may also be a critical, if unacknowledged, driver in some cases. There has been speculation that the reason for so much interest in nuclear energy from states in the Middle East and North Africa is that it may provide a latent nuclear capability to match that of Iran (IISS 2008). Such motivations are likely to be confined to a tiny number of states. In any event, it is not the easiest and fastest way to acquire nuclear weapons (Alger 2009). There is further discussion of proliferation issues in Chapter 2.

Nuclear promotion internationally

Since 2000 the nuclear industry and pro-nuclear energy governments have sensed a second chance for nuclear energy and have themselves become important drivers of interest. Emphasizing all the substantive drivers considered above, with the notable addition of climate change as a new motivator, they have vigorously promoted what they call a nuclear energy 'renaissance'. The three most active countries have been the United States, France and Russia.

In the United States, where the term 'renaissance' was invented, the administration of President George W. Bush was one of the greatest proponents of a nuclear revival, with strong support from industry, the bureaucracy (notably the Department of Energy), Congress and some state governments. After announcing its Nuclear Power 2010 programme in 2002, the administration was active both domestically in supporting nuclear energy and in promoting it internationally, especially through the establishment of the Global Nuclear Energy Partnership (GNEP) (Pomper 2010). Internationally, GNEP has stirred interest in nuclear energy where it might otherwise not have existed, most noticeably in the cases of Ghana and Senegal. The controversial US–India nuclear cooperation agreement, signed in 2005, was promoted as a lucrative commercial opportunity for US businesses to participate in India's ambitious nuclear energy plans – despite the country's status as a nuclear-armed nonparty to the 1968 Nuclear Non-proliferation Treaty (NPT). The United States has also been active in concluding nuclear cooperation agreements with aspirant nuclear energy states, notably the UAE, Jordan and Vietnam. The Obama administration is noticeably less enthusiastic about nuclear energy, has cancelled the domestic part of GNEP and refocused the international programme on non-proliferation objectives.

Pinning its hopes on a global export market for Electricité de France (EDF) and Areva, France has been even more active than the United States, especially in promoting reactor sales. President Nicholas Sarkózy has personally pursued such benefits for his country in a series of international visits during which bilateral nuclear cooperation agreements have been reached, most notably in the Middle East and North Africa. Since 2000 France has signed at least 12 of these[9] and become particularly engaged in promoting the use of nuclear energy by Algeria, Jordan, Libya, Morocco, Tunisia, Qatar and the UAE. Sarkózy has been matched in his efforts by high-profile Areva CEO Anne Lauvergeon, dubbed by the *New York Times*' Roger Cohen as 'Atomic Anne' (Cohen 2008). Areva has already taken advantage of the opening up of the Indian market by signing a contract for up to six reactors (Dow Jones Newswires 2009).

France's aggressive nuclear marketing tactics have drawn criticism, including from then IAEA Director General Mohamed ElBaradei who has warned that they are 'too fast' (Smith and Ferguson 2008). It has also been reported that French national nuclear regulator Andre-Claude Lacoste, who has some say in approving French reactor exports, has suggested to President Sarkózy that he be 'a little bit more pragmatic' about signing nuclear cooperation agreements with countries that currently have no nuclear safety infrastructure (MacLachlan 2008c). NRC Chair Dale Klein has said that as Sarkózy 'goes around the world trying to sell the French reactor, it puts Lacoste in a challenging position in terms of the time it will take for such countries to develop such infrastructure' (MacLachlan 2008c).

The French government has, to its credit, established an international nuclear cooperation agency, L'Agence France Nucléaire International (AFNI), as a unit of the Commissariat à l'Énergie Atomique (Atomic Energy Commission), to 'help foreign countries prepare the institutional, human and technical environment necessary for installation of a civilian nuclear program under conditions of safety, security and nonproliferation' (MacLachlan 2008b). In addition, France's Nuclear Safety Authority (ASN) issued a position paper in June 2008 saying it will impose criteria for cooperation with countries seeking to commence or revive a nuclear power programme, since building the infrastructure to safely operate a nuclear power plant 'takes time', and that it would be selective in providing assistance (Inside NRC 2008: 14).

Russia is the third major promoter of nuclear energy internationally, having reorganized its nuclear industry in 2007 into a vertically integrated holding company, Atomenergoprom, to compete with Areva and other emerging vendors (Pomper 2009). Russia envisages Atomstroyexport (ASE), the nuclear export arm of Atomenergoprom, becoming a 'global player', capturing 20 per cent of the worldwide market and building about 60 foreign reactors within 25 years (Pomper 2009). Past Russian reactor exports have been to Eastern European states, China, India and Iran. Russia has actively pursued nuclear cooperation agreements with other countries in recent years, including some considering nuclear energy programmes, notably Algeria, Armenia, Egypt and Myanmar (Burma) where it is supplying a research reactor. Russia has also signed sales agreements with China, Bulgaria, India, Myanmar and Ukraine, although these do not necessarily relate to reactors but to enrichment and fuel services.

Having always seen spent fuel as a resource rather than waste, Russia enthusiastically embraced GNEP, and in May 2008 signed a nuclear cooperation agreement with the United States to further its bilateral and global nuclear energy activities. The agreement is currently in limbo in the US Congress due to the conflict with Georgia in 2008, although this may be a pretext for Congressional annoyance at advancing civilian nuclear cooperation with a country that has not always been helpful on the Iran nuclear issue (McKeeby 2008). Russia, on the other hand, has also sought to ease US and other countries' proliferation concerns about a nuclear energy revival – and seize commercial advantage – by establishing an International Uranium Enrichment Centre at Angarsk in Siberia, which includes a fuel bank to help provide assurances of supply to countries considering nuclear power. Kazakhstan, Ukraine and Armenia are joining this venture.

International organizations are also playing an international promotional role and to that extent are prominent among the drivers of the revival. The key multi-lateral players are the IAEA and the NEA. The IAEA is constrained in promoting nuclear energy too enthusiastically by its dual mandate, which enjoins it to both advocate the peaceful uses of nuclear energy and help ensure that this occurs safely, securely and in a non-proliferating fashion. Having learned its lesson in over-optimistically forecasting the growth of nuclear energy in the 1980s, the IAEA is today usually more sober in its projections than industry or some of its member states. It also usefully advises new entrants to the nuclear energy business to carefully consider all the requirements for successfully acquiring nuclear energy, notably through its exhaustive *Milestones in the Development of a National Infrastructure for Nuclear Power* (IAEA 2007c).

Outgoing Director General Mohamed ElBaradei claimed that he never preached on behalf of nuclear energy: 'The IAEA says it's a sovereign decision, and we provide all the information a country needs' (BAS 2009: 7). More pointedly, in regard to the current enthusiasm for nuclear energy, he told the *Bulletin of the Atomic Scientists* in an interview in September 2009 that:

> In recent years, a lot of people have talked about a nuclear renaissance, but I've never used that term. Sure, about 50 countries were telling us they wanted nuclear power. But how many of them really would develop a nuclear power program? Countries such as Turkey, Indonesia and Vietnam have been talking about building nuclear power plants for 20 years. So it's one thing to talk about nuclear power: it's another thing to actually move forward with a program.
>
> (BAS 2009: 7)

Nonetheless, the IAEA occasionally becomes overly enthusiastic, such as its claim on its website in July 2009 that 'A total of 60 countries are now considering nuclear power as part of their future energy mix' (IAEA 2009e), a figure apparently derived from a list of countries that had at any time, at any level, approached the agency for information on civilian nuclear energy.

The Nuclear Energy Agency, founded in 1958, also promotes nuclear energy. It is a semi-independent body attached to the OECD, whose membership comprises the most economically developed states,[10] which between them have 85 per cent of the world's installed nuclear energy capacity. In theory the NEA is freer to promote nuclear energy generally and among OECD member states than the IAEA, since its mandate is not complicated by non-proliferation considerations to the same extent. However, as the NEA itself notes, the positions of its member countries regarding nuclear energy 'vary widely from firm commitment to firm opposition to its use' (NEA 2008: 2). The European members are in particular sharply divided. The agency's role is thus supposedly confined to providing 'factual studies and balanced analyses that give our members unbiased material on which they can base informed policy choices' (NEA 2008: 2). In practice, NEA publications often read like a paean to nuclear energy, highlighting the advantages while playing down the disadvantages. The NEA also competes for attention with another part of the OECD, the IEA, which, with its broad energy mandate, has traditionally been less enthusiastic about nuclear.

For industry's part, it has the World Nuclear Association (WNA) as its principal cheerleader. Since its transformation from the fuel-oriented Uranium Institute in 2001, the WNA has attempted to promote the civilian nuclear power industry as a whole worldwide. It naturally has a vested interest in painting the rosiest picture of nuclear energy that it can (Kidd 2008: 66). Yet it faces the challenge of representing a fragmented industry lacking a 'critical mass of strong powerful companies with good public images to stand up for it' (Kidd 2008: 182). Other industry organizations like the World Association of Nuclear Operators (WANO) are explicitly not devoted to promoting nuclear energy but other aspects such as safety and security.

Nuclear promotion domestically

National political pressures have emerged in many countries for revisiting nuclear energy: from public opinion, from within government and from domestic industry. Paradoxically, one of the domestic drivers of a reconsideration of nuclear energy in several countries, especially in Europe and the United States, has been a rise in public acceptance – if not support. This has apparently been stimulated by other drivers, including worries about climate change, economic growth and energy prices, and the availability and long-term security of supplies (MIT 2003: 72). Support is highest in countries with operating nuclear power plants and where residents feel well-informed about radioactive waste issues.

The results of a global public opinion poll, which surveyed 10,000 people online in 20 countries in November 2008, were released in March 2009 by Accenture, a UK-based management consultancy firm. While only 29 per cent supported 'the use or increased use' of nuclear power outright, another 40 per cent said they would change their minds if given more information (Accenture Newsroom 2009). Twenty-nine per cent said they were more supportive of their country increasing the use of nuclear energy than they were three years ago, although 19 per cent said they were less supportive. The top three reasons for opposition were: waste disposal (cited by 91 per cent), safety (90 per cent) and decommissioning (80 per cent) (Nuclear News Flashes 2009a: 2). Demands for improved safety and security have become progressively greater since the nuclear industry began, sometimes to such an extent that, according to the industry, the economics of nuclear have been adversely affected due to delays in approvals, the need for public hearings and constantly evolving safety and other regulatory requirements. Yet the industry has benefited in the past from public input (Smith 2006: 183), and in any case needs to keep public opinion on side by complying fully with regulatory requirements.

In the United States, a Zogby International poll in 2008 revealed that 67 per cent of Americans support the construction of new nuclear power stations (WNN 2008h). Public support is also apparently rising in Britain: over half the respondents in an April 2008 survey felt that the UK should increase its nuclear capacity. Those living closest to existing nuclear plants were most strongly in favour (WNN 2008k). A 2008 survey in Italy, the only country ever to completely abandon an existing nuclear power programme, showed that 54 per cent of respondents were now in favour of new nuclear plants in the country (WNN 2008i). Support appears to run highest in Russia, where a Levada Center poll in 2008 found that 72 per cent of Russians felt that nuclear power should be 'preserved or actively developed' (WNN 2008g).

A European Commission poll of opinion in the European Union in 2007 showed that opinion is divided. Only 20 per cent supported the use of nuclear energy, while 36 per cent had 'balanced views' and 37 per cent were opposed (NEA 2008: 343). A Eurobarometer poll conducted in 2008 showed opinion moving in favour of nuclear: since 2005 support increased from 37 per cent to 44 per cent, while opposition dropped from 55 per cent to 45 per cent (Public Opinion Analysis Sector, European Commission 2008). Forty per cent of opponents would reportedly change their minds if there were a safe, permanent solution to the radioactive waste problem.

However, large sectors of public opinion in many countries remain sceptical about nuclear power and increased support is often conditional and fragile. A poll for the Nuclear Industry Association (NIA) in the UK in November 2007 indicated a falling off from previous increases, a growth in the number of people undecided and 68 per cent of respondents admitting they knew 'just a little' or 'almost nothing' about the nuclear industry (WNN 2007). NIA Chief Executive Keith Parker described the result as a 'reality check' for the industry.

Meanwhile, a study based on five years of research on how local residents view their nearby nuclear reactors has concluded that the 'landscape of beliefs' about nuclear power does not conform to 'simple pro- and anti-nuclear opposites' (Pidgeon *et al.* 2008). It concludes that even among those accepting of a nuclear power station in their midst, support is conditional and could easily be lost if promises about local development of nuclear power are not kept or if there is a major accident anywhere. One also needs to be careful about cause and effect. Rather than representing a deep-felt reconsideration of nuclear energy which is then reflected in public policy, increased public support may be due to changes in politicians' attitudes and increased advocacy of the nuclear alternative by governments. Public support may thus wither with a change of government, as may occur in the UK with the advent of a Conservative/Liberal Democrat coalition government that is internally deeply divided over nuclear energy.

In general therefore, public opinion seemingly is either mildly encouraging, a constraint or in flux rather than a driver, and remains especially preoccupied with the issues of nuclear waste, nuclear safety and security and weapons proliferation. The industry itself is aware that it would only take another serious nuclear reactor accident to kill public support for a nuclear revival.

Domestic support for new nuclear build in many cases may depend crucially on the strength of the pro-nuclear lobby compared with other energy lobbyists and the anti-nuclear movement. American and French companies appear to be best at promoting their industry and seeking government subsidies and other support. Traditionally there has been a close relationship between the US Department of Energy (DOE) and American nuclear companies and utilities, but this may be changing due to concerns about global warming and the role of other types of energy in dealing with it. The DOE's budget was almost entirely devoted to nuclear energy until the Obama administration's recent addition of millions of dollars for alternative energy. Mycle Schneider makes the case that France's nuclear industry has consistently advanced due to the close relationship between government and industry, particularly due to the virtual monopoly of Corps des Mines graduates on key positions (Schneider 2009). Japan, South Korea and Russia are also examples where a close relationship exists between the nuclear industry and government that helps drive promotion of the nuclear industry.

Although the nuclear industry can be a powerful lobby, its influence should not be exaggerated. Steve Kidd notes that a significant problem in encouraging a more positive image of the industry is that it 'isn't really an industry at all, but a separate set of businesses participating in various parts of the nuclear fuel cycle' (Kidd 2008: 66). Some companies have interests in other forms of energy, such as coal, as do utility companies, making it difficult to find strong industrial advocates solely for nuclear energy.

Conclusions

There are several significant drivers of the current interest in nuclear energy, some based on concerns and motivations widely accepted as convincing and legitimate, such as increasing demand for electricity, the pursuit of diversity of supply and the increasingly urgent need to tackle global warming. Yet a number of others are based less on rational analysis and feasible plans than wishful thinking (such as a massive, crash programme of nuclear deployments to deal with climate change). Some come with worrying side-effects (such as increased demand for water in a water-deprived world) or unintended consequences (such as increased rather then decreased energy security). Yet others simply represent the pursuit of vested commercial or national interest regardless of the broader consequences. Taken together, these drivers may, or may not, be sufficient to overwhelm the constraints in the eyes of national policy-makers. In the next chapter the constraints on the future expansion of nuclear power, including the downsides of several of the drivers, will be considered.

2 Assessing a nuclear energy revival

The constraints

Multiple factors are acting as constraints on the expansion of the use of nuclear energy worldwide. The strength and mix of factors varies from country to country, making it often impossible to generalize globally. Many of the factors are matters of contention and controversy, most obviously the question of the economic feasibility of new nuclear build and the challenges of nuclear waste disposition. The fact that these issues generate controversy creates uncertainty for governments, industry and private financiers considering investing in nuclear energy. This uncertainty itself acts as a brake on a nuclear energy revival.

Nuclear economics

According to the NEA, 'Economics is key in decision making for the power sector' (NEA 2008: 173). Promoters and critics of nuclear power are in agreement. The director of strategy and research at the WNA, Steve Kidd, affirms that 'Whether or not nuclear power plants are built and whether they keep operating for many years after commencing operation is these days essentially an economic decision' (Kidd 2008: 189). Nuclear energy critic Brice Smith notes that 'The near-term future of nuclear power ... rests heavily on the predictions for the cost of building and operating the next generation of reactors compared to the cost of competing technologies' (Smith 2006: 29).

Stark disagreement exists regarding the comparative costs of energy alternatives. The IEA in 2008 proclaimed: 'Projected costs of generating electricity show that in many circumstances nuclear energy is competitive against coal and gas generation' (IEA 2008a: 283). Mark Cooper, senior fellow for economic analysis at the Institute for Energy and the Environment at Vermont Law School, concludes: 'Notwithstanding their hope and hype, nuclear reactors are not economically competitive and would require massive subsidies to force them into the supply mix' (Cooper 2009b: 66). The WNA's Director-General John Ritch says, 'In most industrialized countries today, new nuclear power offers the most economical way to generate base-load electricity – even without consideration of the geopolitical and environmental advantages that nuclear energy confers' (WNA 2008c: 4). For Steve Thomas of the Public Services International Research Unit at the University of Greenwich, 'If nuclear power plants are to be built in Britain, it seems clear that extensive government guarantees and subsidies would be required' (Thomas 2005). According to the Washington DC-based lobby group, the Nuclear Energy Institute, 'nuclear power can be competitive with other new sources of baseload power, including coal and natural gas' (NEI 2009: 1).

Assessing the current and future economics of nuclear power, whether on its own, or in comparison with other forms of generating electricity, is complex. This is due, first, to the large number of variables and assumptions that must be taken into account, notably the costs of construction, financing, operations and maintenance (O&M), fuel, waste management and decommissioning. Second, the size and character of the industry make the costs and benefits of government involvement significant and often critical. Such involvement includes: direct and indirect subsidies (sometimes amounting to bailouts); the establishment and maintenance of a regulatory framework to ensure safety, security and non-proliferation goals; and in recent years the possible imposition of a carbon tax and/or greenhouse gases cap-and-trade system that may improve the economics of nuclear energy. A third source of complexity is the lack of recent experience in building nuclear power plants, rendering the real costs of construction and likely construction periods unknowable. According to Joskow and Parsons, the confusion and debate about costs is largely a consequence of 'the lack of reliable contemporary data for the actual construction of real nuclear plants' (Joskow and Parsons 2009: 50). As the NEA has noted, 'These factors are likely to make the financing of new nuclear power plants more complex than in previous periods' (NEA 2008: 203).

Finally, currently there is an unprecedented degree of uncertainty in the energy sector across the board, which in turn affects the economics of nuclear power: climate change considerations are forcing governments to consider all forms of energy production and compare their comparative costs and other advantages and disadvantages; recent wild fluctuations in the price of fossil fuels and other commodities have made predictions of future prices appear less reliable; and the 2008–2009 global financial crisis and resulting global economic slowdown have sharpened investor scrutiny of capital-intensive projects like nuclear.

The (rising) cost of nuclear power plants

Nuclear power plants are large construction undertakings. In absolute terms they are dauntingly expensive. The Olkiluoto-3 1,600 MW EPR currently being built in Finland had a fixed price of €3 billion ($4 billion) in 2003, but is now 50 per cent more than that (WNN 2009g). The UK's Department of Trade and Industry (DTI) used as its central case a reactor cost of £2.8 billion (NEA 2008: 180).[1] In Canada, the quote from AECL for two new CANDU reactors at Darlington, Ontario, was reportedly CAD$26 billion, while Areva's bid came in at CAD$23.6 billion (*Toronto Star* 2009). In 2007, Moody's quoted an 'all-in price' of a new 1,000 MW nuclear power reactor in the United States as ranging from $5–6 billion each.[2]

Lew Hay, chairman and CEO of Florida Power and Light, has noted that 'If our cost estimates are even close to being right, the cost of a two-unit plant will be on the order of magnitude of $13 to $14 billion'. 'That's bigger', he quipped,

> than the total market capitalization of many companies in the U.S. utility industry and 50 percent or more of the market capitalization of all companies in our industry with the exception of Exelon.... This is a huge bet for any CEO to take to his or her board.
>
> (Romm 2008: 4)

The WNA is not sure about whether nuclear is unique or not, asserting that 'Although new nuclear power plants require large capital investment, they are hardly unique by the standards of the overall energy business, where oil platforms and LNG [liquid natural gas] liquefaction facilities also cost many billions of dollars', but then noting that 'Nuclear projects have unique characteristics. They are capital intensive, with very long project schedules' (WNA 2009d: 6, 21).

Since 2003 construction costs for all types of large-scale engineering projects have escalated dramatically (MIT 2009: 6). This has been due to increases in the cost of materials (iron, steel, aluminum and copper), energy costs, increased demand, tight manufacturing capacity and increases in labour costs (IEA 2008b: 152). Yet, not only are costs of nuclear plants large, they have been rising disproportionately. According to the 2009 MIT study update of its 2003 study, the estimated cost of constructing a nuclear power plant has increased by 15 per cent per year heading into the current economic downturn (MIT 2009: 6). The cost of coal- and gas-fired plants has also risen but not by as much. Companies in the United States planning to build nuclear power plants have announced construction costs at least 50 per cent higher than previously expected (IEA 2008b: 152). While some of these costs may currently be falling due to the global economic downturn, they are likely to rebound once the slump reverses and demand from China, India and Japan begins to increase once more. Some price rises are,

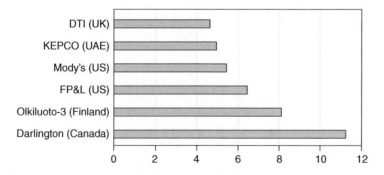

Figure 2.1 Cost estimates of recent nuclear build (per unit) (sources: *World Nuclear News* 2009g; NEA 2008: 80; *Toronto Star* 2009; Moody's Corporate Finance 2007; Romm 2008).

however, unique to nuclear, brought about by shortages of reactor components, notably large forgings for reactor vessels.

Cost comparisons with other baseload power sources

The major traditional competitors with nuclear for 'baseload' power are coal, natural gas and, to a declining extent, oil. Competitive energy markets tend to heighten the disadvantages of nuclear. Coal and natural gas plants are cheaper and quicker to build, they obtain regulatory approval more easily, are more flexible electricity generators (they can be turned on and off easily) and can be of almost any size. Construction takes between two and three years for a combined cycle gas turbine (CCGT) and between one and two years for an open-cycle turbine (IEA 2008b: 143–144). The MIT and University of Chicago studies give lead times of two to three years for natural gas (Smith 2006: 38).[3] All of these factors explain the boom in gas-fired construction in the UK, the United States and elsewhere in recent years, notwithstanding the rising price of fuel (currently receding from the peak). Coal-fired plants can also be built relatively quickly. The MIT and Chicago studies agree on four years' construction time for coal-fired plants, as does the NEA (NEA 2005: 36).[4]

Nuclear plants take up to ten years to plan, obtain regulatory approval for and build, their upfront costs are huge and they are inflexible generators. Essentially, in order to be economic, nuclear power plants need to be as large as possible and kept operating at full power. They cannot readily be shut down and restarted to cope with fluctuations in electricity demand. But light water reactors do need to be shut down periodically to refuel (although CANDU and other heavy-water plants do not).

To calculate the total cost of nuclear energy in order to compare it with other forms of energy production, two concepts are commonly used. The industry tends to use 'overnight costs', the spending on construction materials and labour as if the plant were to be constructed 'overnight', expressed as the cost per kilowatt (kW). The IEA cites studies putting the overnight cost of nuclear power in the United States at $1,500–2,000, in the UK at £1,150–1,250 and in the EU at €1,280–1,900 (IEA 2008a: 290). This hypothetical construct, a 'form of virtual barn raising', does not include the cost of financing the construction and other costs such as escalation of expenses due to increased material and labour costs and inflation (Cooper 2009b: 20). The term 'all-in cost', expressed in the same units as overnight costs, attempts to include these and is thus useful for determining the effects of construction delays (WNA 2008c: 4). But as Mark Cooper points out, 'facilities are not built overnight, in a virtual world' and what utilities and governments wish to know is the cost of electricity that needs to be passed on to the consumer (Cooper 2009b: 20).

Hence the use by economists of a concept known as 'levelized cost' (sometimes known as the 'busbar' cost). This is the minimum price at which a particular technology can produce electricity, generate sufficient revenue to pay all of the costs and provide a sufficient return to investors (CBO 2008: 16).[5] The levelized cost traditionally takes into account the overnight construction cost, plus the costs of financing, fuel, operation and maintenance (O&M), waste disposition and decommissioning. These are calculated for the lifetime of the plant, averaged over that lifetime and expressed as the price of delivered electricity per kilowatt hour (kWh) or megawatt hour (MWh).

After construction costs, the next biggest cost is the money used for the project, whether borrowed or drawn from savings or other funds already held. A 'discount rate' is used, expressed as a percentage figure, to express the value of such money over the time it is used for the project. The rate fluctuates depending on the assessed risk of the project – the higher the estimated risk, the higher the discount rate. The discount rate is presumed to take into account all known risk factors, including political, technical and environmental.

Some caveats about levelized cost models are, however, necessary. First, they produce widely varying results depending on the assumptions made in selecting the input data (particularly for the so-called base case or starting point). Second, the models, which are strictly economic, do not normally take into account all the factors influencing the choices of investors in deregulated electricity markets, notably income taxes and financial conditions (the level of investor confidence in all electricity generation projects).[6] Moreover, the discount rate, although said in theory to encompass 'perfect knowledge' of all risk factors, relies on judgement calls by financial experts about some risks which are not necessarily quantifiable, notably political risks. The discount rate applied to nuclear power plants can vary enormously, as shown in Table 2.1 (from 5 per cent to 12.5 per cent). Finally, the sheer size, expense and unpredictability of nuclear projects may introduce risk factors that cannot be 'internalized' into cost estimates. Economist David McLellan also points out that while the gap between the levelized costs of nuclear and gas may appear narrow, the latter 'should be more attractive to private investors than the costs difference alone suggests', since it requires much less upfront investment and repays the investment more quickly (McLellan 2007: 6). (Some investors, such as pension funds, may however be content with longer-term returns.)

Fortunately, cost estimates for nuclear power are today made by a much wider variety of stakeholders than ever before. While in the 1970s and 1980s boom they were made largely by nuclear vendors, utilities and governments, today these have been joined by independent analysts, academics and investment consultants, including Wall Street firms like Moody's and Standard and Poor's.

As illustrated in Table 2.1, which shows the results of the most prominent recent studies on the economics of nuclear power, even accounting for currency conversion difficulties the ranges in cost estimates are enormous. The overnight cost per kilowatt ranges from $690 to more than $3,000, while the generating cost ranges from $15 per MWh to $78 per MWh. The variance illustrates the complexity of the decisions facing potential investors in nuclear energy, which are considered below.

In 2003 the Massachusetts Institute of Technology (MIT) published what is probably the most widely cited study on the future of nuclear power. Although

Table 2.1 Results of recent studies on the cost of nuclear power

	Year	Original currency	Cost of capital	Overnight cost (per kW)		Generating cost (per MWh)	
				Original	2000 USD	Original	2000 USD
Massachusetts Institute of Technology (MIT)	2003	USD	11.5%	2,000	1,869	67	63
Tarjamme and Luostarinen	2003	EUR	5.0%	1,900	1,923	24	25
Canadian Energy Research Institute	2004	CAD	8.0%	2,347	1,376	53	31
General Directorate for Energy and Raw Materials, France	2004	EUR	8.0%	1,280	1,298	28	28
Royal Academy of Engineering	2004	GBP	7.5%	1,150	725	23	15
University of Chicago	2004	USD	12.5%	1,500	1,362	51	46
IEA/NEA (High)	2005	USD	10.0%	3,432	3,006	50	41
IEA/NEA (Low)	2005	USD	10.0%	1,089	954	30	25
Department of Trade and Industry, UK (DTI)	2007	GBP	10.0%	1,250	565	38	18
Keystone Center (High)	2007	USD	11.5%	4,000	3,316	95	89
Keystone Center (Low)	2007	USD	11.5%	3,600	2,984	68	63
MIT Study Update	2009	USD	11.5%	4,000	3,228	84	78

Source: adapted from IEA (2008b: 290). Historical exchange rates and GDP deflator figures adapted from US GPO (2009a, 2009b).

having a strong US policy emphasis, this multidisciplinary study also produced findings that it implied were applicable worldwide. The MIT group began from the premise that the need to reduce greenhouse gas emissions to tackle climate change was so great that re-evaluating the role of nuclear energy was justified. In that sense the report was predisposed towards a nuclear revival. Unlike the IEA/NEA, which in its calculations applied the same discount rate to all baseload forms of electricity generation, the MIT group applied a higher weighted cost of capital (essentially the discount rate) to the construction of a new nuclear plant (10 per cent) than for coal or natural gas plants (7.8 per cent) (MIT 2009: 8). The result of its calculations was that nuclear was likely to be more expensive than coal and CCGT, even at high natural gas prices. At such prices for nuclear and gas, it was coal rather than nuclear that would attract new plant investment. The report bluntly declared: 'Today, nuclear power is not an economically competitive choice' (MIT 2003: 3).

Convinced that nuclear energy should still make a contribution to the energy mix in tackling global warming, the MIT researchers suggested that progressive achievement of cost reductions in the nuclear industry (by reducing construction costs, construction times and O&M costs and by securing financing on the same terms as gas and coal) could make it comparable in price with coal and gas. This assumed moderate prices for coal and gas, although gas is often cheap. The authors judged – in 2003 – that such cost improvements by the nuclear industry, while 'plausible', were not yet 'proven'. It added that nuclear would become more competitive if the 'social cost of carbon emissions is internalized, for example through a carbon tax or an equivalent "cap and trade" system' (MIT 2003: 7). It also advocated government financial incentives for a few 'first-of-a-kind' new entrants, to 'demonstrate to the public, political leaders and investors the technical performance, cost and environmental acceptability of the technology' (MIT 2009: 19).

In a 2009 update, the MIT group expressed disappointment that six years later the economics of nuclear remained essentially unchanged (MIT 2009: 6). While the price of natural gas had risen dramatically, making nuclear more attractive (although gas has since retreated from its peak), the construction costs for all types of large-scale engineering projects had also escalated dramatically – but more so in the case of nuclear. MIT's estimated overnight cost of nuclear power had doubled from $2,000/kW to $4,000/kW in six years.

The 2009 update noted that:

> While the US nuclear industry has continued to demonstrate improved operating performance [for existing reactors], there remains significant uncertainty about the capital costs, and the cost of its financing, which are the main components of the cost of electricity from new nuclear power plants.
>
> (MIT 2009: 6)

It suggested, unsurprisingly, that lowering or eliminating the risk premium would make a significant contribution to making nuclear more competitive.

Table 2.2 Costs of electricity generation alternatives

	Overnight cost ($/kW)	*Fuel cost ($/mmBtu)*	*Levelized Cost Of Energy (LCOE)*		
			Base case (¢/kWh)	*w/carbon charge $25/tCO₂ (¢/kWh)*	*w/same cost of capital (¢/kWh)*
2003*					
Nuclear	2,000	0.47	6.7		5.5
Coal	1,300	1.20	4.3	6.4	
Gas	500	3.5	4.1	5.1	
2009**					
Nuclear	4,000	0.67	8.4		6.6
Coal	2,300	2.60	6.2	8.3	
Gas	850	7.00	6.5	7.4	

Source: adapted from MIT (2009: 3). See source for original assumptions on which the calculations are based.

Notes
* Based on 2002 USD; ** Based on 2007 USD

This will only occur, it said, through 'demonstrated performance' by 'first movers' which will in turn only occur because of government subsidies to lower the risk.

The 2005 US Energy Policy Act (EPA) – which the 2003 MIT study may have helped inspire – provided such government support. Yet four years later the MIT researchers judged that the programme had not been effective in stimulating new build. This may be in part because, as the MIT researchers themselves had advocated, all low-carbon technology alternatives have been subsidized. This confirms Mark Cooper's claim that technology neutral subsidies do 'not change the consumer economics much' (Cooper 2009b: 61).

The 2007 Keystone Center's Nuclear Power Dialogue, involving 11 organizations, nine of which were corporations or utilities involved in selling or buying nuclear power plants, estimated that the life-cycle levelized cost of future nuclear power in the United States would have a 'reasonable' range of between 8 and 11 cents per kWh delivered to the grid (Keystone Center 2007: 11). This is higher than MIT's 7.7–9.1 cents per kWh (in 2007 dollars). The study took into account a likely range of assumptions on the critical cost factors, such as escalation in material costs, length of construction period and capacity factor. It noted that while some companies have announced their intention to build 'merchant' plants (those owned and operated by companies which then sell their electricity to utilities or other distribution companies), 'it will be likely be easier to finance nuclear power in states where the costs are included in the rate base with a regulated return on equity' (Keystone Center 2007: 12).

Since it is investors, both private and public, whose money would be at risk in investing in nuclear power, it is instructive to consider their views. In short, private capital remains sceptical, as do utilities being pressed to invest in new build. In October 2007, Moody's Investors Service concluded that 'the ultimate costs associated with building new nuclear generation do not exist today' and that 'the current cost estimates represent best estimates, which are subject to change' (Moody's Corporate Finance 2007). It estimated that the overnight cost of a new nuclear power plant in the United States could range from \$5,000 to \$6,000 per kW. For the new build in Ontario, Canada, Moody's is forecasting an overnight cost of \$7,500 per kW (compared to the Ontario Power Authority's claim of \$2,900 per kW) (Ontario Clean Air Alliance 2000). As for levelized costs, according to Mark Cooper, numerous studies by Wall Street and independent energy analysts estimate efficiency and renewable costs at an average of six cents/kWh, while nuclear electricity is in the range of 12–20 cents/kWh (Cooper 2009b: 1).

Cost comparisons with non-baseload alternatives: conservation, efficiency and renewables

One argument used for increasing the use of nuclear energy is that no other relatively carbon-free alternatives exist for providing 'reliable baseload power', especially for large urban areas. Although the term is often misused, baseload power in the parlance of the electricity industry means power at the lowest operating cost. This baseload is supplemented during peak periods by costlier forms of generation. Mark Cooper claims that utilities promote a narrow focus on traditional central power station options since 'large base load is what they know and they profit by increasing the base rate' (Cooper 2009b: 43). Yet renewable energy like wind and solar could be run as baseload power since the operating costs are marginal compared with large power plants.

Baseload power also does not mean 'most reliable', since all sources of electricity are unreliable to varying degrees and power grids are designed to cope with highly variable supply. Nuclear power plants must be shut down periodically for refuelling and maintenance, prolonged heat waves may deprive them of cooling water as occurred in France in 2003, and they also automatically shut down for safety reasons during electricity blackouts, at the very time they are most needed, as occurred in the great northeast US blackout in August 2003 and in Brazil in November 2009 (Gordon 2009). In fact, a portfolio of many smaller units is inherently more reliable than one large unit as it is unlikely that many units will fail simultaneously. Moreover, at least in the United States, 98–99 per cent of power failures originate in the grid rather than in the power generation source (Lovins and Sheikh 2008: 24). As Socolow and Glaser put it,

> The inflexibility of base load nuclear power and the intermittency of wind and solar share the feature that neither of these low CO_2 emitters can meet a time-varying demand for electric power without assistance from complementary systems: load following and peaking plants and storage.
>
> (Socolow and Glaser 2009: 35)

While most baseload power is projected to continue to come from centralized power stations, simply because they already exist and already supply major portions of total electricity demand, there are many cheaper alternatives. At the very least nuclear will find itself competing in terms of both investment and subsidies, as governments seek to adjust their energy mix for reasons of climate change and energy diversity (IEA 2008b: 45).[7]

Some of these alternatives would be able to reduce demand for baseload power, such as conservation and efficiency measures. In the IEA's ACT and BLUE scenarios to 2050, energy-efficiency improvements in buildings, appliances, transport, industry and power generation represent the largest and least costly savings (IEA 2008b: 40). One example is combined heat and power (CHP), which by generating electricity and heat simultaneously, can increase the overall efficiency of electricity and heat generation and reduce the environmental footprint. For existing industrial facilities or power stations using biomass, natural gas or coal (but not nuclear), only a modest increase in investment costs is required. Despite a loss of efficiency in electricity generation (increasing heat production hurts electricity production) (MacKay 2009: 146, 149), this can be 'very profitable' according to the IEA (IEA 2008b: 143). As Socolow and Glaser note, 'In a world focused on climate change mitigation, one would expect massive global investment in energy efficiency – more efficient motors, compressors, lighting, and circuit boards – that by 2050 could cut total electricity demand in half, relative to business as usual' (2009: 35)

Cost comparisons between nuclear and such alternatives can be even more complex than comparing baseload alternatives. Yet, as Mark Cooper points out, compared to the diversity of nuclear cost estimates, there is much less diversity in the cost estimates of alternatives, so the figures tend to be more convincing. In a comparison of six recent studies, Cooper reveals that 'New nuclear reactors are estimated to be substantially more expensive than a variety of alternatives, including biomass, wind, geothermal, landfill, and some solar and conventional fossil fuels' (Cooper 2009b: 43).

Naturally, the levelized cost is only one aspect of the choices policy-makers will face in choosing their national energy mix. As David McLellan confirms, 'The economics of nuclear plants vary from one country to another, depending upon energy resource endowments, government policies and other factors that are country specific' (McLellan 2007: 18).

It is beyond the scope of this study to analyse all of the pros and cons of these technologies and the likelihood that they could supplant significant amounts of baseload power generation to the point of persuading policy-makers to forego new nuclear builds. Clearly, many alternative energy sources face significant challenges, including the intermittency of supply (wind and solar); the need for enormous tracts of land in order to generate sufficient amounts of energy (wind, solar, biofuels) and energy storage capacity (battery technology). Other technologies, such as 'clean coal' and CCS are unproven and subject to great scepticism. However, R&D is proceeding at a pace for some technologies whereby improvements in performance and cost will likely arrive more quickly than they can for nuclear technology – which has demonstrated long lead times, poor technology (or industrial) learning rates and large cost-overruns.

The IEA notes that the likelihood of 'technology learning' occurring is an important factor in making decisions about whether to invest in emerging energy technologies. 'Technology learning' refers to the notion that over time the costs of a new technology would be expected to fall as a result of experience gained during the manufacturing and deployment process (this does not include cost reductions achieved through a deliberate R&D process). The IEA, which has surveyed observed historic learning rates for various electricity supply technologies (IEA 2008a: 205) has concluded that the learning rate for nuclear in the period 1975–1993 in the OECD area was just 5.8 per cent, the lowest except for offshore wind and CCS. By comparison, onshore wind achieved learning rates of between 8 and 32 per cent, while photovoltaics ranged from 20–23 per cent. A study by McDonald and Schrattenholzer shows the learning rates of the nuclear industry compared to other selected energy technologies to be low (6 per cent compared to 17 per cent for wind, 32 per cent for solar and 34 per cent for gas turbine combined cycle (GTCC)) (McDonald and Schrattenholzer 2001: 355–361). Observed learning rates for various demand-side technologies were all higher than nuclear, including selective window coatings (17 per cent), facades with insulation (17–21 per cent), double-glazed coated windows (12–17 per cent) and heat pumps (24–30 per cent). While such comparisons need to be treated with care due to the difficulty of comparing different technologies, especially those starting from a low base (like wind) compared with more mature ones like nuclear, it does give an indication of the competition for investment in R&D and deployment that nuclear faces.

An additional challenge for nuclear, beyond cost and industrial learning rates, is that Thomas Edison's twentieth-century model of electricity production and distribution, with large centralized generating plants and long-distance delivery by gigantic electricity networks, is being rethought in ways that are not favourable to nuclear power.[8] The new model involves 'smart grids' that can better balance supply and demand (*The Economist* 2009b: 71), combined with 'distributed generation' from small plants closer to the customer. Unless the nuclear industry can produce small, cheap, safe and secure nuclear plants located near the customer, this new model favours alternative energy sources, such as solar, wind and biofuels, for baseload and off-peak power (IEA 2008b: 143), as well as encouraging energy conservation.

The effects of deregulation of electricity markets

Unlike the initial boom in nuclear plant construction in the 1970s and 1980s, many countries' energy markets have today been deregulated or partially deregulated. Although no country has a purely private electricity market, a competitive situation now exists in most OECD countries and several non-OECD ones (IEA 2008b: 155). This has made realistic assessments of the economics of all forms of energy, including nuclear, more important. Private investors and electricity utilities are today much more likely than in the past to base their decisions to invest in nuclear power on its projected cost compared to other forms of generating electricity and the likely rate of return on their investment compared to the alternatives.

When markets were regulated, utilities were not required to bear the full risk of investment in nuclear power plants. Instead, they employed reactor techno-logy that had been developed by governments, often as by-products of a nuclear weapons programme, and the costs of which had been written off. In addition, reactor companies and utilities were directly or indirectly subsidized by govern-ments to build nuclear power plants. And finally, utilities were able to pass on costs to the consumer without fear of being undercut by competition.

Today, as the 2003 MIT study notes, 'Nuclear power will succeed in the long run only if it has lower costs than competing technologies, especially as electric-ity markets become progressively less subject to regulation in many parts of the world' (MIT 2003: 7). No new nuclear power plant has yet been built and oper-ated in a liberalized electricity system. The first attempt, in Finland, is experienc-ing significant cost overruns. The WNA's Steve Kidd calls the delays in the Finnish project a 'significant blow' to demonstrating the viability of new build in deregulated markets (Kidd 2008: 50).

Even in un-deregulated markets the economics are proving challenging. In China, with the most grandiose plans for nuclear energy and where it might be thought that public funding is no barrier, the economics are apparently of concern. The head of the China Atomic Energy Authority, Sun Qin, has explained that once China's nuclear power plants are operating the power is competitive (especially since a fixed, low electricity price is guaranteed for all types of power), but 'we must resolve the problem of initial investment' (Nuclear News Flashes 2007a). This is despite the fact that government directives allow Chinese utilities to invest 'spectacular amounts in capital equipment, financed largely by cheap loans from state-controlled banks' (*The Economist* 2010b: 72). In un-deregulated Turkey, where by law the state guarantees to pay the plant owner-operator a fixed price for electricity, nuclear plans were at one stage derailed when the unit price was deemed too high for the Turkish state to guarantee.

Construction delays and cost overruns

Major one-off engineering and construction mega projects like bridges, tunnels and Olympic stadiums almost invariably take longer to build and cost more than originally estimated. In the nuclear industry delays and cost overruns are legion. According to the World Energy Council, the average construction time for nuclear plants increased from 66 months in the mid-1970s to 116 months (nearly ten years) between 1995 and 2000 (Thomas *et al.* 2007: 10).[9] Since 2000 there has been a decline, but average construction time remains at seven years (Thomas *et al.* 2007: 10). Nuclear plant construction projects are so capital intensive, so complex and of such duration that even relatively minor delays can result in significant cost overruns. Mark Cooper notes that 'Reactor design is complex, site-specific, and non-standardized. In extremely large, complex projects that are dependent on sequential and complementary activities, delays tend to turn into interruptions' (Cooper 2009b: 41). The challenge for the nuclear industry is that, unlike other one-off projects, such as bridges and stadiums, nuclear power plants must compete with cheaper alternatives.

Part of the rationale for Generation III+ reactors is the hope that costs will come down with 'learning experience' from subsidized first-of-a-kind reactors, economies of scale from multiple new builds, modularization and assembly-line production of components (Schneider 2009: 32),[10] as well as 'advanced project management techniques' (Boone 2009: 8–9).[11] (In addition, the industry is pinning its hopes on streamlined government regulation, as in the UK and United States). However, as we have seen, the learning that usually lowers costs over time has not generally occurred in the nuclear power business. A study by Mark Cooper demonstrates that during the 'great bandwagon' era of American nuclear build in the 1960s and 1970s 'on average, the final cohort of ... market reactors cost seven times as much as the cost projection for the first reactor' (Cooper 2009b: 2). Although the IEA estimates that a learning rate of just 3 per cent is required to reduce its current estimated cost of Generation III+ nuclear power plants from $2,600/kW in 2010 to a 'commercialization target' of $2,100/kW by 2025, this seems grossly unrealistic, especially since the MIT 2009 update gives a current overnight cost for new reactors in the United States of $4,000/kW (IEA 2008a: 206). (For Generation IV reactors the IEA figures are even less convincing: a 5 per cent learning rate was said to be needed to bring the estimated cost of $2,500/kW in 2030 to a post-2050 figure of $2,000/kW.)

Faster construction times are currently being recorded in Asia. Of the 18 units built in Asia between 2001 and 2007, three were connected to the grid in 48 months or less (IEA 2008a: 287). The fastest was Onagawa 3, a Japanese 800 MW boiling water reactor (BWR) connected in 2002 after a 41-month construction period (IEA 2008a: 287). According to AECL, its CANDU-6 reactors being built in China were delivered ahead of schedule and under budget (Oberth 2009). In fact, AECL claims that in the last 13 years it has contractually delivered seven reactors on time and on budget in China, South Korea and Romania. The South Koreans claim a construction time of 50 months for its APR1400 (KEPCO 2009).

These may be exceptions that prove the rule, since the current nuclear 'revival' is, in Mark Cooper's words, showing 'eerie' parallels to the 1970s and 1980s. Cooper claims that 'startlingly low-cost estimates prepared between 2001 and 2004 by vendors and academics and supported by government officials helped create what has come to be known as the "nuclear renaissance"' (Cooper 2009b: 2). Two top GE Hitachi executives, Jack Fuller and Danny Roderick, confirm that reactor vendors did the whole industry a 'disservice' in the past by announcing optimistic costs that later proved to be unattainable' (MacLachlan 2009c: 1). According to Cooper's research, the most recent cost projections for new nuclear reactors are, on average, more than four times as high as the initial 'nuclear renaissance' projections (Cooper 2009b: 1). This has not prevented nuclear boosterism of the type that characterized the 1970s and 1980s from re-emerging. According to Steve Kidd,

> What is needed is the courage to get over the initial period of pain of high initial capital costs to enter the 'land of milk and honey' in subsequent years, where nuclear plants can be almost 'money machines' for their owners.
>
> (Kidd 2008: 203–204)

Such cost overruns are not restricted to the United States. India's reactors have all been over budget, ranging between 176 and 396 per cent (Thomas *et al.* 2007: 11). In Canada, the Darlington facility built in the period 1981–1993 so compromised the financial position of the provincial utility, Hydro Ontario, that its CAD$38 billion debt was orphaned into a separate fund. Provincial electricity consumers to this day see an amount added to their electricity bill to pay off this debt (Cadham 2009: 6).

The EPR Areva reactor currently being built in Finland at Olkiluoto is turning out to exemplify what might go wrong with a nuclear energy revival. Problems have arisen in the Olkiluoto-3 project despite the long, largely positive experience of Finland with nuclear power in the past and its reputation as an efficient, well-governed and highly regulated state, and despite the reputation of Areva as a leading nuclear company. As of November 2009 the project was more than three years behind schedule (the plant is expected to open 'beyond' June 2012 (TVO 2009)) and was more than 50 per cent (€1.7 billion) over budget (WNN

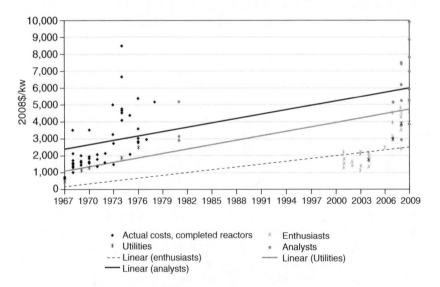

Figure 2.2 Overnight cost of US pressurized water reactors: actual cost of completed reactors and projections (with sources of projections identified) (source: Cooper 2010: v).

2009g). GE Hitachi senior vice-president Danny Roderick blamed the problems on insufficient design detail when construction began, as well as Areva's continued use of 'stick-built' or, on-site, construction techniques dating back to the 1970s (MacLachlan 2009c: 1). This seems to call into question claims that this first-of-a-kind reactor will lead to production line cost savings for future EPRs.

There are several lessons to be learned from the Olkiluoto-3 project for the nuclear revival:

- Turnkey contracts represent a huge risk for plant vendors, which in the future are likely to be wary of them (this has already occurred in Ontario, where the bids were starkly realistic and thus too expensive for the government to contemplate).
- New nuclear build may not be economically viable even in partially deregulated markets.
- The skills needed to successfully build a nuclear plant to the engineering and safety standards required are considerable and a lack of recent experience of such construction may make the challenge even greater.
- There are serious challenges for regulatory bodies, even those as professional and experienced as the Finnish Radiation and Nuclear Safety Authority (*Säteilyturvakeskus* or STUK), in overseeing new generation reactor construction, especially first-of-a-kind plants (STUK had not assessed a new reactor order for more than 30 years) (Thomas *et al.* 2007: 40).

The impact of the current financial crisis

The 2008–2009 financial crisis and global economic recession have added to existing investor uncertainty about the economic fundamentals of nuclear energy. An IEA background paper for the G8 energy ministers meeting in May 2009 reported that energy investment worldwide is plunging in the face of a tougher financial environment, weakening 'final' demand for energy and falling cash flows (IEA 2009a). It estimated that global electricity consumption could have dropped by as much as 3.5 per cent in 2009 – the first annual contraction since the Second World War. The IEA expects to see a shift to coal- and gas-fired plants at the expense of more capital-intensive options such as nuclear and renewables, although it added that 'this will depend on the policies and support mechanisms individual countries and regions have in place' (IEA 2009a: 4). Platts reported in September 2009 that according to industry leaders at the WNA annual symposium in London 'the international financial and economic crisis that began a year ago has cast a chill over the burgeoning nuclear "renaissance"' (MacLachlan 2009b: 1–3).

Meanwhile, the US Nuclear Energy Institute (NEI) has dramatically scaled back its Vision 2020 plan launched in 2002 to foster the addition of new US plants by 2020. It now projects only 20,000 MW compared to 50,000 MW. Marvin Fertel, NEI president and CEO, said this was due to the current economy and the absence of new units demonstrating that they could be successfully built

(Weil 2009b: 4). Plans for new build by the government and utility Bruce Power in Ontario, Canada, have been cancelled due to both falling demand for electricity and rising costs. South Africa has done likewise.

Although governments around the world have sought to attenuate the effects of the recession by pumping government funds into their economies, notably by supporting infrastructure projects, there is no evidence that governments anywhere have chosen nuclear power projects to stimulate quick economic growth. Nuclear projects are unlikely to have the desired 'shovel-readiness' and, given the long lead-times and specialized skills required, are unlikely to produce instant jobs as part of an economic stimulus package.

The impact of government subsidies

Governments, like utilities and the nuclear industry, will take economic and financial considerations into account in making public policy decisions about whether or not to permit, support, actively encourage or actually invest in nuclear energy. Some governments, with large reserves derived from oil or other wealth, may simply choose to build nuclear power plants regardless of the economics of nuclear power, but most cannot afford this luxury. Naturally, industry will take into account the willingness of governments to provide a favourable investment environment for nuclear power.

The nuclear energy industry has always been the beneficiary of government financial support, either direct or indirect. This is attributable to several factors: the technology emerged from nuclear weapons or other military programmes; the nature of the industry requires significant regulation and oversight by governments (in terms of safety, security, non-proliferation and waste); governments needed to assume insurance liability above certain limits in case of catastrophic accidents to encourage nuclear build to proceed; many of the earliest reactor projects lacked commercial viability; and construction delays and cost overruns forced governments to absorb or retire unmanageable debt incurred by utilities. Governments in states with national reactor vendor companies also willingly became financially involved in promoting what was seen in the 1960s and 1970s as a revolutionary technology that promised huge profits from exports. The most famous example is the US Atoms for Peace programme, but other governments, like Canada, with its support of CANDU reactor exports, joined in (Bratt 2006).

The nuclear industry is currently schizophrenic on the issue of subsidies, trumpeting the 'new economics' of third-generation nuclear reactors, while seeking government assistance to kick-start a revival. This includes preferential treatment for 'first-of-a-kind' plants in the hope that they will lead to a flood of orders and production-line techniques that will lower costs. According to Kidd, 'The first new nuclear units to be built should not now need financial subsidies, as the economics now look sound, assuming that investors can take a long-term view' (Kidd 2008: 79), although 'Initial plants of new designs ... face substantial first-of-a-kind engineering costs and may need some public assistance to become economic' (Kidd 2008: 44).

American studies all conclude that for the United States, at least, the most critical factor in the future relative cost of different electricity generation technologies is government financial support (along with the price of carbon) (CBO 2008: 26).[12] The United States seems to offer the greatest variety of open subsidy mechanisms through its 2005 Energy Policy Act (EPAct), including loan guarantees, tax credits, regulatory delay insurance and other subsidies for the first handful of new reactors, as well as funding from the Department of Energy for 'first-of-a-kind' reactors (CBO 2008: 11, 8–9).[13] President Obama announced in February 2010 the first loan guarantee, $8.3 billion for the Southern Company to build two new nuclear reactors at its Vogtle generating station near Augusta, Georgia (Spotts 2010). These would be the first new nuclear reactors built in the United States in 30 years.

The Congressional Budget Office concludes, however, that: 'EPAct incentives by themselves could make advanced nuclear reactors a competitive technology for *limited additions* [emphasis added] to baseload capacity' (CBO 2008: 2). The 2003 MIT study, on the other hand, concludes, despite the magnitude of the subsidies they propose, that these would still not be large enough to fully overcome the higher costs of nuclear compared with fossil fuels: this would only be done through a carbon tax and by reducing the risk premium through 'demonstrated performance' (MIT 2009: 8). Under the MIT proposal the levelized cost of electricity from the ten 'first movers' would be approximately 6.2 cents per kWh, which is still well above its estimate for the price of electricity from coal or natural gas (Smith 2006: 49).

The 2009 update of the 2003 MIT study, which had advocated 'limited government assistance' for 'first mover' US nuclear plants, concludes that this has 'not yet been effective in moving utilities to make firm reactor construction commitments' (MIT 2009: 9). This was due to three reasons: the DOE has not moved expeditiously enough to issue the regulations and implement the programme; the requirement of many state governments that utilities obtain a certain fraction of their electricity from low-carbon sources has excluded nuclear; and increased cost estimates are making the industry seek even more assistance. The NEI has argued that the current loan guarantee programme of $18.5 billion is 'clearly inadequate' (Alexander 2009) and proposed at least $100 billion for all clean energy technologies, including nuclear (WNN 2009e). The MIT group opposes increased subsidies, arguing for a level playing field for all energy generation technologies based on a carbon tax or a cap-and-trade system (MIT 2009: 10). Despite this the Obama administration has proposed tripling the amount available for loan guarantees from the $18.5 billion allocated in 2005 (Wald 2010).

According to Mark Cooper: 'Seeking to override the verdict of the marketplace, the industry's lobbying arm has demanded massive increases in subsidies from taxpayers and ratepayers to underwrite the industry', but even with subsidies nuclear would still be more expensive than the alternatives (Cooper 2009a: 1). Peter Bradford, a member of the Nuclear Regulatory Commission from 1977 to 1982, argues that 'the US can revert to the sensible notion of limited support for a few first mover nuclear projects or it can insist that US taxpayers continue to underwrite a "revival" that the industry has proven unable to manage' (Bradford 2009: 64). As the Congressional Budget Office (CBO) warns, 'under some plausible assumptions … in particular those that project higher future construction costs for nuclear plants or lower gas prices – nuclear technology would be a relatively expensive source of capacity, regardless of EPAct incentives' (CBO 2008: 2). The CBO considers the risk of default on the part of the nuclear industry to be very high – well above 50 per cent (CBO 2003).

The UK, meanwhile, has said it will take active steps to 'open up the way' to construction of new nuclear power stations, but conscious that British taxpayers have borne the costs of past failures to achieve nuclear energy profitability (Brown 2008: 3)[14] has made clear that it is up to private enterprise to fund, develop and build the new stations (Davis 2009: 26). Ontario, meanwhile, has signalled it is not prepared to subsidize nuclear by guaranteeing cost overruns.

Currently, many governments engaged in the 'revival' are offering 'support' for new build, although not necessarily in the form of subsidies. In most of these countries, including China, India, Russia, Japan, Taiwan, South Korea and the Ukraine, a lack of transparency about costs and hidden subsidies makes it impossible to ascertain the complete extent of such support. Even in Western market economies information about costs and subsidies may be difficult to discern. In France, the true total cost of nuclear energy, closely linked as it is to the nuclear weapons fuel cycle, is apparently considered a state secret and has never been disclosed (Brown 2008: 32).[15] Some countries with oil wealth such as Nigeria, Saudi Arabia, the UAE and Venezuela, may be able to provide the ultimate in government subsidy by simply buying a nuclear plant outright with government funds, without transparency and little or no legislative oversight. Self-financing of nuclear plants by China enables it, according to *The Economist*, to do away with the 'uncertainty and delay of negotiating guarantees with international development outfits or bilateral export-credit agencies' (*The Economist* 2010b: 72).

Yet most governments, burned by past experience with nuclear cost overruns and construction delays, constrained by deregulated markets, facing demands for a level playing field for different energy technologies and strapped for cash in the current economic downturn, will remain reluctant to provide subsidies.

The impact of carbon pricing

The CBO concludes that 'the longer-term competitiveness of nuclear technology as a source of electricity is likely to depend on policy makers' decisions regarding carbon dioxide constraints' (CBO 2008: 26) that could increase the cost of generating electricity with fossil fuels. The 2003 MIT study, too, noted that while nuclear power was not currently economically competitive, it could become so if future carbon dioxide emissions carried a 'significant price' (MIT 2003: 8). The CBO notes that the effect is most pronounced for coal, which emits nearly a metric ton of carbon dioxide for every megawatt hour of electricity produced (CBO 2008: 2). Even modern coal- and gas-fired plants designed to use fuel more efficiently would still emit enough carbon dioxide to make nuclear more economic under a carbon regime. Plants that use CCS, which has not yet been proven commercially, are likely to emit just 10 per cent of the carbon dioxide of current fossil fuel plants. Yet this still fails to compete with the zero emissions from a nuclear generating plant (Galbraith 2009).

Cooper points out, however, that imposing a price on carbon makes all low carbon options, including efficiency and renewables, more attractive. He contends that this would thus 'not change the order in which the options enter the mix' (Cooper 2009b: 8). Brice Smith notes that an increased focus on efficiency as a result of a carbon price would result in reduced demand for electricity, throwing into question the need for large power plants (Smith 2006: 60). The World Business Council for Sustainable Development contends that some alternative technologies like ultra-supercritical pulverized coal (USSPC) and wind in optimal locations are already 'mature' and would be competitive were the value of CO_2 emissions internalized into electricity prices (WBCSD 2008: 3).

Crucially, carbon taxes and/or cap-and-trade systems rely on private enterprise and investors to respond to market signals. It could be a decade before the price of carbon stabilizes at high enough levels for confident investment decisions to be made about using nuclear energy instead of other sources. The European Union's pioneering system, established in 2005, while understandably fraught with teething problems, has still not priced carbon high enough for nuclear to become economical (Nuclear News Flashes 2009b: 3).[16] In 2009 the MIT group lamented that a carbon tax, along with other incentives for nuclear had not yet been realized, meaning that 'if more is not done, nuclear power will diminish as a practical and timely option for deployment at a scale that would constitute a material contribution to climate change risk mitigation' (MIT 2009: 4). The system enacted by the US Congress in June 2009 has been watered down by making early distribution of permits largely free, ensuring that politics rather than sound economics will govern the price, at least in its early years.

Essentially the prospects of a global price on carbon are so uncertain as to make it impossible for investors today to assess the effects on the economics of nuclear power. The whole future of the international climate regime is itself in doubt, especially after the failure of the Copenhagen climate change conference in December 2009. Implementation of a global price for carbon through either a tax or a cap-and-trade system is years away, while investment decisions about nuclear energy need to be made now.

Costs of nuclear waste management and decommissioning

The costs of nuclear waste management and decommissioning of civilian power reactors should ideally be 'internalized' into the cost of electricity. Contrary to popular perception, long-term waste management and decommissioning are a negligible part of the overall estimated costs of nuclear, since they are calculated in future dollar values which, due to inflation, become progressively cheaper.

Thomas *et al.* note that if a 15 per cent discount rate for a new power plant is applied to decommissioning and waste management they essentially 'disappear' from the calculations (Thomas *et al.* 2007: 60). However, they also claim that it would be wrong to apply such a high rate of return to such long-term liabilities since funds collected from consumers should be placed in low-risk investments to minimize the possibility that they will be lost. Such investments yield a low interest rate (Thomas *et al.* 2007: 60).

A more pertinent question is whether future costs of waste management and decommissioning are adequately estimated, especially given the fact that 'no full-size nuclear power plant that has completed a significant number of years of service has ever been fully dismantled and disposed of' (Thomas *et al.* 2007: 45). Moreover, there is no experience anywhere in the world with long-term disposal of high-level nuclear waste from civilian nuclear power plants (although the United States does have such experience with high-level military waste).

Ideally, funds for nuclear waste management and decommissioning should accumulate from revenues obtained from the electricity generated. Such funds may either be held or managed by the commercial operator (as in France and Germany) or by the government (as in Finland, Sweden and the United States). Problems may arise if utilities are unwilling or unable to set aside real funds (as opposed to a notional, bookkeeping entry); if funds are lost through poor investments; or if the company collapses before the end of a plant's expected lifetime. All of these have occurred in the UK, where significant decommissioning costs for old nuclear plants (estimated in 2006 at around £75 billion and rising) will be paid by future taxpayers since real funds were not set aside (Thomas *et al.* 2007: 26–27, 60). The NEA estimates, alarmingly, that in some cases in the EU existing funds set aside for decommissioning represent less than 50 per cent of the anticipated real costs, although it reports that steps are being taken to redress this situation (NEA 2008: 265). In the United States the government is being sued by nuclear utilities for collecting monies for centralized nuclear waste management but failing to provide it at Yucca Mountain. The NEA pleads for decommissioning funds to be 'sufficient, available and transparently managed' (NEA 2008: 265), something neither the nuclear industry nor governments have achieved to date.

Industrial bottlenecks

After finance, a second major constraint on rapid expansion of nuclear energy is said to be a lack of industrial capacity. Since the first major expansion of nuclear energy in the 1980s, capacity specific to building nuclear power plants has until recently atrophied everywhere, except in France, Japan and South Korea. Arguments have been advanced that, globally, industry would not be able to sustain a major nuclear energy revival, at least to 2030, because the scale of activity required is unprecedented.

In the 1980s approximately 150 reactors were under construction simultaneously (NEA 2008: 318). In the peak years of 1985 and 1986, 33 of these were connected to the grid. In the 1980s as a whole, an average of one reactor was added every 17 days, but mostly in only three countries (France, Japan and the United States) (NEA 2008: 316). According to the NEA, extrapolation of this historical experience, taken together with the growth in the global economy since that time, suggests that the capability to construct 35–60 1,000 MWe reactors per year could be rebuilt if necessary (NEA 2008: 316). As of April 2010 there were approximately 57 reactors 'under construction' worldwide (IAEA 2010c), so the construction of 60 new ones per year is, in theory, achievable. However some of these units have been under construction for years or were substantially already built and work is resuming to finish them, so the comparison is not precise. Moreover, the NEA notes that in 2030–2050 a much higher build rate will be required to keep pace with the retirement of most existing plants – along with further capacity expansion.

The Keystone Center's 2007 Dialogue agreed that 'the most aggressive level of historic capacity growth (20 GWe/yr) could be achieved or exceeded in the future'. However, this would depend on realizing the claims for advanced reactors: larger output per plant (10–50 per cent), advanced construction methods, greater use of modularization, advances in information management and 'a more competent global supply base' (Keystone Center 2007: 26), some of which are problematic. The Keystone report also provided a useful reminder that not just nuclear plants would need to be built, but in addition, globally:

- 11–22 additional large enrichment plants to supplement the existing 17;
- 18 additional fuel fabrication plants to supplement the existing 24;
- ten nuclear waste repositories the size of the statutory capacity of Yucca Mountain in the United States, each of which would store approximately 70,000 tons of spent fuel.

Keystone participants were unable to reach consensus about the rate of expansion for nuclear power in the world or in the United States over the next 50 years. Some thought it was unlikely that nuclear capacity would expand appreciably above its current levels and could decline; others thought that it could expand rapidly enough 'to fill a substantial portion of a carbon-stabilization "wedge" during the next 50 years' (Keystone Center 2007: 10).

The rate at which countries can ramp up a nuclear energy programme will vary. The United States has a particularly flexible economy that responds quickly to market opportunities, but other market economies, including some in the European Union, such as Italy and former Eastern European bloc countries, are considered less nimble. Semi-command economies with heavy governmental control, like those of China and Russia, may be better able to direct resources where needed.

Since no single company can build a complete nuclear power plant by itself, one challenge is to rebuild what the NEA calls global supply chains, involving numerous contractors and sub-contractors, each of which must achieve the high manufacturing and construction standards required for a nuclear plant extending well beyond the nuclear reactor itself.

The most commonly cited industrial bottleneck relates to ultra-large nuclear forgings used in large nuclear reactor vessels, essentially for units of 1,100 MWe capacity and beyond.[17] Currently the three suppliers of heavy forgings are Japan Steel Works (JSW), China First Heavy Industries and Russia's OMZ Izhora (Kidd 2009a: 10). New capacity is being built in Japan by JSW; in South Korea by Doosan; and in France by Areva at Le Creusot. There are plans for new capacity in the UK by Sheffield Forgemasters and India by Larsen & Toubro. Nothing is planned for North America. Industry is therefore attempting to ramp up in response to demand, as would be expected. However, manufacturers will only respond as long as firm orders are in the pipeline, since investment in major forges and steelmaking lines is not cheap. The difficulty faced by new large nuclear forging entrants is the length of time it takes to gain the necessary technical quality certification, such as that issued by the American Society of Mechanical Engineers (ASME) (Birtles 2009: 42).

Areva's situation is illustrative. Because EPRs are being built in several countries, Areva has to deal with both the French manufacturing code and the ASME code. Guillaume Dureau, head of Areva's equipment business unit, has warned that rather than producing standardized, interchangeable forgings, 'we will have to know what power plant it's for before starting to pour the forgings' (*Nucleonics Week* 2009a: 3–4). This would appear to attenuate one of the advertised benefits of Generation III+ reactors – standardization. He noted that a combination of larger component size, new designs and stricter safety requirements, coupled with the need for more forged components than previous reactor models, had posed huge challenges for Areva's components plant, but 'We have clearly shown we know how to do this' (*Nucleonics Week* 2009a: 3).

Nonetheless, those interested in putting money into the nuclear industry, and the industry itself, face a classic investment catch-22: both will have to be convinced of the likelihood of a major revival before investing in the necessarily specialized and expensive production capacity that would make such a revival possible.

Personnel constraints

The nuclear industry's stagnation since the early 1980s has led to a dramatic decline in enrolment in nuclear science and engineering degrees worldwide, leading to what is now referred to as the 'missing generation'. The OECD/NEA published a report in July 2000, *Nuclear Education and Training: Cause for Concern?*, which quantified, for the first time, the status of nuclear education in OECD member countries (NEA 2000). It confirmed that in most OECD countries nuclear education had declined to the point that expertise and competence in core nuclear technologies were becoming increasingly difficult to sustain. The problems included:

- the decreasing number and dilution of nuclear programmes;
- declining numbers of students taking nuclear subjects and the significant proportion of nuclear graduates not entering the nuclear industry;

- the lack of young faculty to replace ageing and retiring faculty;
- Ageing research facilities, which are being closed and not replaced.

There has also been a significant long-term reduction in government funding of nuclear research in some countries since the mid-1980s, notably in Germany, the UK and the United States – although France and Japan have held up comparatively well (NEA 2008: 322–323).

The existing nuclear workforce is also declining. The nuclear industry, in a steady state over the last few decades, has had relatively little turnover in employees, leading to an ageing workforce and little recruitment. A large portion of the existing nuclear labour force is set to retire within the next five to ten years, including numbers as high as 40 per cent in France and the United States (Squassoni 2009a: 46–47). A shortage of experienced nuclear plant operators – many of the current operators have spent their entire professional lives in the nuclear industry, beginning in the 1960s and 1970s – is particularly troubling. In a 2007 North American Electric Reliability (NERC) survey, 67 per cent of respondents said it is highly likely that an ageing workforce and lack of skilled workers could affect electricity reliability (Schmitt 2008: 3). The impact is likely to be more pronounced for nuclear power because of the special training, experience and licensing criteria required for employment. Such experience cannot be gained simply from training courses. In addition, new skills and competencies will be needed in the new generation technologies envisaged, as well as in decommissioning and nuclear waste management. The skills shortage has been exacerbated by deregulation of electricity markets, which has led to cost-cutting by downsizing the workforce, a loss of research facilities and cuts in financial support to universities (NEA 2004: 8).

On a global basis a rapid short-term expansion in nuclear energy is thus likely to be limited by the shortage of qualified personnel. As the NEA notes, 'It is likely to take several years to redevelop the capability to construct new nuclear power plants, while maintaining the necessary high standards and the ability to keep projects on time and to cost' (NEA 2008: 316). The effects will be felt differently from country to country. For the UK, for example, 'It is clear that the envisaged new nuclear build programme … will be almost like establishing a new industry' (Kidd 2008: 55). Argentina has had to turn to Canada for expertise in resuming construction on its Atucha-II reactor after 14 years – not only had the technology changed, but the personnel had moved on (WNN 2008e). In addition to the industry itself, regulatory bodies and the IAEA will also be competing for experienced personnel.

It will be especially difficult for states, mostly developing countries that are contemplating building a nuclear energy sector from scratch, to attract qualified personnel to build, operate, maintain and regulate their nascent nuclear facilities. Only the wealthiest, such as the oil-rich states, will be able to afford to pay the high salaries in such a competitive market. The UAE has already made a name for itself by siphoning highly trained personnel from other companies and organizations to oversee its nuclear development; as a result, it is one of the more likely aspiring nuclear states to succeed in its plans. Indonesia, Jordan and Vietnam are unlikely to be able to compete on the same terms.

Ramping up educational and training programmes is a long-term project, but a report by the NEA in 2004 noted that steps have been taken by some governments to ameliorate the problem. Efforts have been made in the past decade in the United States, Japan and Europe in particular to increase university enrolment in nuclear science and engineering (Elston 2009). A European Nuclear Education Network (ENEN) has been established to foster high-level nuclear education (NEA 2008: 325–327). The UK has launched a National Skills Academy for Nuclear (NSAN) to coordinate recruitment and training of personnel, while the universities of Manchester and Lancaster are expanding nuclear research and education (WNN 2008d). Canada has a University Network of Excellence in Nuclear Engineering. At the undergraduate level enrolment in nuclear engineering degrees in US universities has increased from approximately 225 students in 1998 to just under 350 students in 2006, although the number of doctoral engineering degrees has steadily declined (NEA 2008: 36–37). The World Nuclear University is a recent initiative taken by the WNA, with participation from the IAEA, the NEA and WANO (WNU 2010).[18] The NEA notes, however, that some OECD governments have not taken any initiatives at all, perhaps because they prefer the private sector to take the lead, because there is a national moratorium on nuclear power or because they consider adequate programmes already exist.

Ultimately, the extent of personnel shortages will depend on the size and scope of the nuclear revival as determined by its other drivers and constraints, and by the agility of both governments and the private sector in responding. While skills deficits may constrain a significant nuclear revival, governments have the capacity to overcome them if they prioritize skills development and training – a decision that will be based on their own predictions about the future of nuclear energy.

Nuclear waste

The final major constraint on a global expansion of nuclear energy is the abiding controversy over radioactive nuclear waste disposal. Not only is it controversial among the general public, being among the most often cited reason for opposing nuclear power, but industry itself is concerned. Excelon, the largest US nuclear utility, has said it has 'serious reservations' about proceeding with new nuclear plant construction until the used fuel management issue has been resolved (Nuclear News Flashes 2009c). The risk to human beings from radioactivity is well known but poorly understood by the general public. In addition to the risk of inhalation and ingestion, radiation can pose 'external' risks to human beings from simply being in proximity to it. The key aim is therefore to concentrate and contain nuclear materials and isolate them from the environment for as long as they remain hazardous.

While nuclear power plants and other nuclear fuel cycle facilities invariably release a small amount of radioactivity directly into the environment, increasingly tight controls, according to the NEA, have led to a 'remarkable reduction in the amount of radioactive effluents' (NEA 2008: 242). The bigger problem today is what to do with nuclear waste.

Nuclear power generates radioactive waste containing a variety of substances having half-lives as short as fractions of seconds to as long as millions of years. Such waste is classified according to the level and nature of its radioactivity — low-level waste (LLW), intermediate-level waste (ILW) and high-level waste (HLW). It is also categorized according to its half-life, whether short-lived (SL) or long-lived (LL).

Low- and intermediate-level waste is produced at all stages of the nuclear fuel cycle, from uranium mining to decommissioning of facilities. Although together this waste represents the greatest volume, it contains only a small fraction of the total radioactivity. For the purposes of disposal, low- and intermediate-level waste is sometimes dealt with together. Short-lived low- and intermediate-level waste is disposed of in simple near-surface landfills or in 'more elaborately engineered' near-surface facilities. Most countries with a major nuclear power programme operate such facilities.

Most of the radioactivity from the nuclear fuel cycle, but the smallest amount by volume, is in spent fuel from nuclear reactors or in high-level waste from reprocessing. By far the greatest challenge is what to do with high-level waste from nuclear reactors and long-lived intermediate waste from reprocessing (often known as transuranic waste).

Responsibility for managing the radioactive waste produced at a nuclear power plant initially lies with the facility operator. The waste is usually stored on-site in water-filled cooling ponds to allow short-lived radioactivity to disappear and heat to dissipate. In some countries, such as the Netherlands, Sweden and Switzerland, spent fuel is moved after several years of storage at the reactor site to a centralized national storage facility. If spent fuel is to be reprocessed it will be transported, after cooling, to a reprocessing facility where the recyclable material (95 per cent of the mass) is separated from what then becomes a high-level waste stream. This is usually stored in vitrified form either at the reprocessing plant or in purpose-built facilities.

Storage of such materials at nuclear facilities is regarded only as an interim management solution as it relies on continued active control and maintenance and is vulnerable to extreme natural events such as earthquakes or fire, and malevolent attacks by terrorists or saboteurs or even attempts at seizure. Interim storage areas in some countries, such as Japan, are rapidly filling up, making a permanent solution imperative. From the beginning of the nuclear age, the nuclear industry had expected that governments would move quickly to provide a long-term solution to the nuclear waste problem. Almost six decades after commercial nuclear electricity was first produced, not a single government has succeeded in doing so.

The principal proposed long-term solution is deep geological burial, involving the emplacement of packaged waste in cavities excavated in a suitable rock formation some hundreds of metres below the surface. According to the NEA, the safety principle is that the rock will provide isolation and containment of the radioactivity to allow for sufficient decay so that any eventual release back to the surface will be at levels comparable to that of natural rock formations and 'insignificant in terms of potential effects on health and the environment' (NEA 2008: 249). This principle is compromised somewhat by a demand by some in the nuclear industry that the 'waste' be retrievable if and when technology permits the fuel in it to be used (Tucker 2009), a concept parodied as envisaging a 'deep plutonium mine'. Retrievability raises questions about whether illicit retrieval might be possible, as well as the cost of burying a resource that may be dug up and used in the future.

The world's only operating deep geological repository for radioactive waste, the Waste Isolation Plant at Carlsbad, New Mexico, was developed specifically for disposal of transuranic waste from the US nuclear weapons programme, but it has demonstrated the feasibility of the concept. Currently, only Finland and Sweden (*Nuclear Engineering International* 2007: 18–20) are well advanced and could have their repositories operating by 2020. According to the NEA, they are expected to be followed by France and Belgium, then Germany, Japan, Switzerland and the UK in the 2030s and 2040s. Several other countries have repositories planned, but have announced no implementation dates before 2050. Others have research and development programmes.

For new entrants into the nuclear power business, with just one or a small number of reactors, establishing their own nuclear waste repositories is likely to be completely unrealistic on the grounds of cost and need. International cooperation is likely to be necessary among the smaller new entrants, although there is great sensitivity in all countries, with the apparent exception of Russia, about acting as a nuclear waste dump for others.

Although there is a virtual consensus among scientists that a long-term geological repository for such nuclear waste is a technically and environmentally sound solution, finding a suitable location for such a repository has proven to be a highly volatile political issue in most states, and has been cited as a major reason for opposition to nuclear power. As the NEA cautions: 'the time necessary from a primarily technical point of view to move from deciding on a policy of geological disposal to the start of waste emplacement operations could be of the order of 30 years' (NEA 2008: 252). This does not take into consideration political and economic barriers which may often be the most daunting. The long lead times, as in the case of Yucca Mountain, provide great opportunity for opposition to develop.

A solution to long-term storage of nuclear waste in the United States has been elusive despite more than two decades of efforts to open a repository at Yucca Mountain in Nevada (Vandenbosch and Vandenbosch 2007).[19] The 1982 Nuclear Waste Policy Act (NWPA) required that the federal government open a waste repository by 1998 to store all of the waste generated by the US civilian and military nuclear programmes. The idea of siting it at Yucca Mountain drew fierce political opposition from Nevada Senator Harry Reid, who campaigned successfully against it for more than 20 years. After the DOE budget for Yucca Mountain was cut annually during the George W. Bush administration (Nuclear Fuel Cycle 2008), President Obama fulfilled his campaign promise to scrap the project in February 2009 (WNN 2009f). Future US policy awaits the recommendations of a panel set up by the Obama administration in January 2010 (Goldenberg 2010).

An evolving approach, pioneered by Sweden and Canada, is to undertake a comprehensive, national consultation process aimed at securing agreement on a long-term nuclear waste management strategy. In Canada's case, a three-year study, emphasizing 'citizen engagement', was undertaken by a specially established Nuclear Waste Management Organization (NWMO). It proposed a policy of Adaptive Phased Management which committed 'this generation of Canadians to take the first steps now to manage the used nuclear fuel we have created' (NWMD 2005: 45). The policy promotes 'sequential and collaborative decision making, providing the flexibility to adapt to experience and societal and technological change' (Dowdeswell 2005). Ultimately, though, it envisages 'centralized containment and isolation of used nuclear fuel deep underground in suitable rock formations, with continuous monitoring and opportunity for retrievability' (Dowdeswell 2005).

Canada's deliberative, democratic process, taking into account the 'ethical and social domains as well as the technical questions' is unlikely to be easily emulated in other states, especially those with a strong anti-nuclear movement or those with undemocratic systems. It is not clear, therefore, that the nuclear industry will be able to turn public opinion around in most countries.

One difficulty is the historic link between nuclear weapons programmes and nuclear energy. Weapons programmes produced far more nuclear waste than civilian industry and were often undertaken as crash programmes with scant regard for public safety or the environment. The massive cleanup of the Hanford nuclear site in Washington state is costing American taxpayers an estimated $200 billion and is scheduled to last for decades (Vandenbosch and Vandenbosch 2007: 119). High-level waste had been stored in 177 tanks, of which 149 were single-walled, 67 of which leaked approximately 100 million gallons of radioactive waste into the subsoil and groundwater. While it is unfair to compare the practices of the 1940s with today's more safety- and environmentally-conscious nuclear industry, the legacy of the weapons linkage lingers and affects public attitudes.

A second difficulty for planned new build in existing nuclear energy states is the existing stockpiles of civilian nuclear waste that are a legacy of past procrastination by governments about disposal. Industry projections of a huge increase in the number of nuclear power plants as part of a revival creates the impression of huge increases in the amount of nuclear waste, even though, according to the NEA, 'historic' spent fuel will continue to dominate the worldwide inventory to 2050, even with a significant revival (NEA 2008: 260). The volume of additional nuclear waste will also pale in comparison to waste volumes from nuclear's continuing biggest rival, coal. As David MacKay points out, whereas the ash from ten UK coal-fired power stations equals 40 litres per person per year, the nuclear waste from Britain's ten nuclear power stations equals just 0.84 litres per person per year, most of which is low-level (MacKay 2009: 169). Although there is growing disenchantment with 'dirty coal' and increasing demands that it be phased out due to its massive contribution to global warming, this is unlikely to particularly benefit nuclear power in the debate over waste since nuclear's other main competitor, natural gas, produces no solid waste.

A third and related difficulty for the nuclear industry is communicating the concept of relative risk to the public. In the United States, coal ash from coal-fired stations, for example, exposes the public to more radiation than nuclear power plants (McBride *et al.* 1978). Yet the public is almost completely unaware of this fact. The WNA's Steve Kidd calls the nuclear industry's handling of the waste issue a 'mess' and an 'own goal', and recommends that the industry never admit that it produces waste 'unless you really have to', adding another 'own goal' by confessing that 'Perhaps this is morally not a completely defensible position, but it makes sound business sense' (Kidd 2008: 155). This illustrates a core issue for the nuclear industry – regaining public confidence through trans-parent, honest engagement over the nuclear waste issue. As Canada's NWMO reported of its public engagement process in Canada 'trust must be built before proceeding with any approach for the long-term management of used nuclear fuel' (NWMO 2005: 75).

Safety, security and weapons proliferation

Among the additional constraints on the advance of nuclear energy most com-monly cited are concerns about the safety and security of nuclear facilities and materials, and the potential for nuclear energy programmes to advance the pro-liferation of nuclear weapons. These play into public opinion as well as being the concern of opinion leaders, policy-makers, governments and international organizations. These issues are considered in Chapters 4, 5 and 6 relating to the global governance of safety, security and non-proliferation.

Special barriers for aspiring nuclear energy states

New entrants to the nuclear energy business face particular barriers, in addition to those outlined above, that reinforce scepticism about the likelihood of a signi-ficant nuclear energy expansion among states that currently do not have nuclear power. The biggest barriers are to be found in the areas of governance, infra-structure and finance.

The IAEA states that it can take at least 10–15 years for a state with no nuclear experience to prepare itself for hosting its first nuclear power plant (IAEA 2007c). For states with a 'little developed technical base the implementa-tion of the first [nuclear power plant] would, on average, take about 15 years' (IAEA 2007a).

Governance

A country's ability to lay the foundations for a nuclear power programme depends on its capacity to successfully manage or at least oversee large and complex projects, and its ability to attract or train qualified personnel. Not least of the requirements is an effective nuclear regulatory infrastructure, and a safety and security system and culture. These are not built overnight. Such capacities must be in place well in advance of construction of the first installation. Many aspiring nuclear energy states have shown that they struggle with managing any large investment or infrastructure projects, for reasons ranging from political corruption to terrorism. Establishing the regulatory infrastructure and safety culture for a nuclear power plant even over a ten-year period poses an immense challenge.

Fortunately, the IAEA and responsible vendors, for the sake of their own reputations, will only assist with new build in states that are able to prove they can safely and securely operate a nuclear facility. Less responsible vendors, eager to break into a market long controlled by a small number of corporations, may be more willing to overlook the need for stringent governance requirements or will give them only superficial consideration.

Physical infrastructure

A major barrier to aspiring nuclear states, especially those in the developing world, is having the physical infrastructure to support a nuclear power plant or plants. This includes an adequate electrical grid, roads, a transportation system and a safe and secure site. The IAEA's 'milestones' document includes a comprehensive list of hundreds of infrastructure 'targets' – including physical infrastructure – for aspiring nuclear states to meet before they should commission a nuclear plant. This includes supporting power generators, a large water supply and waste management facilities (IAEA 2007c). Meeting all of the targets will be a major challenge for most developing aspirant states, requiring them to invest billions of dollars on infrastructure upgrades for several years.

A significant and perhaps surprising constraining factor in terms of infrastructure – and a key measurable of a country's eligibility for nuclear power – is a suitably large or appropriate electricity generating capacity. The IAEA recommends that a single nuclear power plant should represent no more than 5–10 per cent of the total installed generating capacity of a national electricity grid (IAEA 2007c: 39). The WNA claims the number is 15 per cent (WNA 2009a). Taking the IAEA's high estimate of 10 per cent as a median, a state would need to already have an electrical grid with an existing capacity of 9,000 MW in order to support a single 1,000 MW nuclear power plant, or else plan to have it built well before bringing a nuclear reactor online. Even large developed countries with an unevenly distributed population like Canada face such problems. (The Canadian province of Saskatchewan, which is considering nuclear power, has only one million people and a total installed generating capacity of 3,878 MW.)

The main reason for this grid capacity requirement is that if a large power plant represents too great a proportion of grid capacity, it risks destabilizing the grid when it goes offline, either in planned or emergency shutdowns (Schewe 2007: 117). In addition, a reliable independent power source is necessary for the construction and safe operation of a nuclear power plant. An incident in Sweden illustrates the importance of the latter. A loss of offsite power for Sweden's Forsmark 1 reactor in July 2006 handicapped the control room functions and deprived operators of information, making it more difficult to shut down the reactor (MacLachlan 2007a: 10). Lennart Carlsson of the Swedish Nuclear Power Inspectorate said the incident showed that 'modern power supply equipment is sensitive to grid disturbances and they are complex' (McLachlan 2007a: 10). The fact that a sophisticated country like Sweden, with decades of experience, is just discovering this fact should give new entrants pause.

One possible solution for such states is to share a nuclear reactor with regional neighbours to spread the investment risk and to distribute the electricity generated in a larger grid, or to sell excess electricity from a nationally owned reactor to neighbours with a shared grid system. However, national electricity grids tend not to be internationally integrated, so sharing electricity from a jointly owned nuclear power plant would usually require additional investment in grid extension and connection.

Financial indicators

Another main indicator that a state may not be able to follow through with its nuclear plans is its ability to finance a nuclear plant. For relatively poor countries, paying for a nuclear power plant is a massive hurdle, even if the costs are spread out over several years. There is no precise way to measure whether or not a country can afford a nuclear power plant, especially since decisions may be driven by politics, national pride, energy security, industrialization strategy, or, in the worst case, nuclear weapons 'hedging', rather than sound financial analysis or a rational national energy strategy. While stretching a national budget to accommodate a nuclear power plant purchase may be in theory possible, this always implies 'opportunity costs' – what might have otherwise been purchased, especially in the vital energy sector. The challenge of measuring financing ability is further complicated by the diversity of public-private economies among aspiring nuclear energy states. Where private capital is unable or unwilling to invest in nuclear energy development on financial grounds, governments may be willing to do so.

A country's gross domestic product (GDP) is one crude indicator of 'affordability'. States with both a low GDP and a poor credit rating are unlikely to be able to secure a loan for nuclear energy development. This is especially true of states with no credit rating, indicating that there is little outside interest in investing in them at all, much less in a major, inherently risky, infrastructure project.

Procuring loans from international lending institutions such as the World Bank or the Asian Development Bank (ADB) is not an option. These lending institutions do not fund nuclear power projects because, in their estimation, the costs are too often underestimated, they have high upfront capital costs and nuclear reactors are too large and inflexible as electricity generators, particularly for developing countries (World Bank Environment Department 1994: 83–89). According to the World Bank, the possibility of nuclear accidents and nuclear waste that may lead to 'involuntary exposure' of civilians to harmful radiation may have 'environmental costs [that] are high enough to rule out nuclear power even if it were otherwise economic' (World Bank Environment Department 1994: 83–89). The only nuclear project it has ever funded was in Italy in the 1950s (World Bank 2003). While the World Bank has recently acknowledged that nuclear power can contribute to ameliorating climate change, it has not altered its lending policy.

The ADB, for its part, reaffirmed in a June 2009 policy update that it would not fund nuclear projects: 'In view of concerns related to procurement limitations, availability of bilateral financing, proliferation risks, fuel availability, and environmental and safety concerns, ADB will maintain its current policy of non-involvement in the financing of nuclear power generation' (ADB 2009: 32). Other regional lending institutions – including the Inter-American Development Bank (IDB), the Islamic Development Bank (IsDB) and the Arab Bank for Economic Development in Africa (BADEA) – do not have nearly enough financial resources to make a meaningful contribution to a nuclear power reactor project.

Export/import credit agencies established by governments may assist with finance in order to boost their domestic reactor manufacturers or governments may provide foreign assistance to cover part of the cost. Canada has done both in promoting CANDU exports to developing countries (Bratt 2006: 79).[20] France and Russia may do so in the future. Of course, states may seek multiple funding sources to spread the risk, including a combination of government finance, commercial loans and foreign aid.

Conclusion

While there is no scientific way of calculating the balance of drivers and constraints acting on such a complex phenomenon as the purported nuclear energy revival, it is hard to avoid the conclusion that the constraints, at least as currently calculable, outweigh the drivers. In particular this study should elicit scepticism about linear projections of increased nuclear energy based on population increases, economic growth rates or electricity demand, sometimes combined with unquestioning extrapolations of governments' announced plans. These methods invariably ignore or discount political, financial and societal factors. Governments tend to exaggerate their nuclear energy expansion plans for political purposes, yet all face political, economic, technological and/or environmental challenges to their ambitions, sometimes including outright anti-nuclear sentiment that needs to be factored into any assessment. As this chapter makes clear, the economics of nuclear energy, compared with those for alternative energy sources, both traditional like coal and natural gas, and newly emerging alternatives such as wind and solar, as well as conservation and efficiency, would appear to be particularly unimpressive. Naturally, these conclusions derive from an assumption that governments, industry and other stakeholders will act rationally when confronted with the pros and cons, an assumption that cannot always be sustained. The next chapter will consider more closely the likely nuclear energy revival in the coming decades, both globally and on the basis both of categories of states and individual ones.

3   Assessing the likelihood of a revival

This chapter considers the official projections that are being made about the growth of nuclear energy to 2030, the extent of the revival so far, and the national plans that states have announced. Taking into account the various drivers and constraints examined in the previous chapters the likelihood that such plans will be realized will be examined. Since drivers and constraints will affect different types of states in different ways, consideration will be given first to the states that currently have nuclear energy and seek to expand their capacities. This will be followed by an examination of the 'aspirant' nuclear energy states and the particular barriers that they face in acquiring nuclear energy for the first time.

Global growth projections to 2030

The IAEA, the most authoritative international source of information on nuclear energy, predicted in August 2009, as its high scenario, a doubling of global nuclear power capacity by 2030, from the current 372 gigawatts electric (GWe)[1] to 807 GWe. The projection assumed an end to the present financial crisis, continued economic growth and electricity demand, and the implementation of policies targeted at mitigating climate change (IAEA 2008e: 6–7).[2] The IAEA's low scenario for the same period projected an increase to just 511 GWe, reflecting a 'conservative but plausible' revival.

The NEA, in its first ever *Nuclear Energy Outlook*, released in 2008, projected a total of just under 600 GWe by 2030 as its high scenario, while its low scenario indicated only a negligible increase over the current level, with new plants built only to replace old ones (NEA 2008: 19, 27). This put both its high and low scenarios below the IAEA's. The NEA's study of various 'business as usual' energy scenarios devised by other international organizations concludes that by 2030 and even by 2050 'fossil fuels (coal, oil and natural gas) will provide a growing share of energy supply, while nuclear power will not make a significant contribution to meeting demand growth' (NEA 2008: 94–95). 'Business as usual' includes no significant effort to tackle carbon emissions through a carbon tax or 'cap-and-trade' system and no effort to promote (and presumably subsidize) nuclear energy.

The IEA, traditionally more sceptical about nuclear energy and with a much broader energy mandate than its fellow OECD agency the NEA, predicted in its 2008 nuclear reference scenario that world nuclear capacity would rise to just 433 GWe by 2030. This puts its estimate considerably below the IAEA's low estimate and even below that of the NEA (NEA 2008: 148). Although nuclear electricity output is expected to increase in absolute terms in all major regions except OECD Europe, with the largest increases occurring in Asia, the IEA assumes that by 2030 nuclear's share of global electricity production will have fallen from 15 per cent in 2006 to 10 per cent, 'reflecting the assumption of unchanging policies towards nuclear power' (NEA 2008: 142–143). In comparison, coal's share of total world electricity production is projected to grow from 41 per cent currently to 44 per cent by 2030, 85 per cent of the increase coming from China and India. Oil is expected to drop to just 2 per cent. Gas demand is expected to drop due to higher prices, leaving its percentage share slightly lower by 2030 at 20 per cent, but new plants, using high-efficiency gas turbine technology, will mostly meet the bulk of incremental gas demand. Since 2008 falling gas prices and the discovery of new methods of extracting gas from shale will have affected this projection. While plants with carbon capture and storage are likely to make only a minor contribution to electricity generation by 2030, the share of renewables is likely to rise considerably, from 18 per cent in 2006 to 23 per cent by 2030. In its 2009 survey the IEA reassessment of China's nuclear energy plan has led it to predict 10 per cent higher global nuclear capacity by 2030 than its previous projections (IEA 2009b: 98).

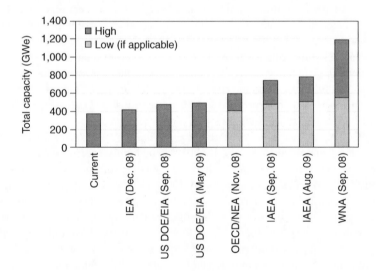

Figure 3.1 Global nuclear energy projections at 2030 (sources: IAEA 2008e: 17; Energy Information Administration 2008: 233; Energy Information Administration 2009: 251; NEA 2008: 19; WNA 2009b).

The WNA, meanwhile, postulated in its 2008 *Nuclear Century Outlook* a high scenario of 1,203 GWe for 2030 and low scenario of 552 GWe, both considerably higher than the IAEA's projection.[3] It describes these not as 'growth scenarios as such, but rather the boundaries of a domain of likely nuclear growth' (WNA 2009b). The US Energy Information Agency (EIA), part of the US Department of Energy, predicted in September 2008 that, 'despite considerable uncertainty about the future of nuclear power', world nuclear generating capacity would rise to 498 GWe in 2030, slightly higher than the IAEA's low scenario, but 31 per cent higher than the IAEA's 2003 projection (EIA 2008). This would be the equivalent, it said, of adding approximately 124 new 1,000 MW reactors[4] to the current world reactor fleet of approximately 436 reactors of varying capacities (EIA 2008).

The 2003 multidisciplinary study by the Massachusetts Institute of Technology (MIT) projected 1,000 GWe of operating nuclear power globally by 2050 (MIT 2003: ix). Six years later its 2009 update estimates that this is 'less likely' than it seemed in 2003 (MIT 2009: 5).

As even a cursory examination indicates, projections of a global nuclear revival are highly variable and not necessarily predictive. First, they are often built on extrapolations of national economic growth, demand for energy generally and demand for electricity in particular which are added together to produce global trends. None of these national indicators necessarily translates specifically into demand for electricity generated by nuclear energy. The WNA *Outlook*, for instance, is 'built on country-by-country assessments of the ultimate growth potential of national nuclear programs, based on estimates of need and capability with projected population a key factor' (WNN 2008c). Four of the scenario sets used by the NEA for its 2008 *Nuclear Energy Outlook*, the exception being those of the IAEA, used computer-based energy modelling incorporating such assumptions (NEA 2008: 92–93). As the recent economic downturn indicates, such 'guesstimates' may be ill founded. As the IAEA concedes of its own figures, these 'should be viewed as very general growth trends whose validity must constantly be subjected to critical review' (IAEA 2008e: 5) As the NEA points out, a larger role for nuclear depends crucially on government policies (NEA 2008: 100).

A second reason for scepticism about nuclear energy projections is this very fact – that aggregate figures are usually derived from totalling governments' announced policies and plans, in what the IAEA calls a 'bottom up approach' (IAEA 2007b: 3). An exception is the IEA, which says its figures reflect 'the consistency of our rule not to anticipate changes in national policies – notwithstanding a recent revival of interest in nuclear power' (IEA 2008b: 39). Current plans by governments, as in the past, are often overly optimistic, designed for internal political consumption and/or to impress neighbouring states, their immediate region or even international bodies like the IAEA. In some regions, notably Southeast Asia and the Gulf, there has been discernible competition between states to be the first to acquire such 'modern' technological artefacts. This is not a new phenomenon, but has characterized the history of nuclear energy from the time of the Atoms for Peace programme in the 1950s and 1960s onwards (Pilat 2007).

In compiling their aggregate global data, international organizations, especially those whose members are governments, have difficulty refuting national estimates, notably those derived from political ambition rather than fact-based analysis. The IAEA's current low projection, for instance, is based on the assumption that 'all nuclear capacity currently under construction or in the development pipeline gets constructed and government policies, such as phase-outs, remain unchanged' (IAEA 2008i). Its high scenario is based on 'government and corporate announcements about longer term plans for nuclear investments, as well as potential new national policies, such as responses to international environmental agreements to combat climate change'. Essentially, these scenarios assume 'full implementation of the long-term plans announced by governments and power utilities' (IAEA 2007b: 3). To its credit, the Agency's estimates are not prepared in-house, but are established by an expert consultancy on Nuclear Capacity Projections (IAEA 2008e: 6). Yet they are ultimately reliant on information supplied by member states. Hence their projections are often overinflated and are never exceeded by reality, as shown in Figure 3.2.[5]

Some past projections have been wrong not just because of governments' over-optimism about future projects, but because of cancellations of projects already underway. Figure 3.3 shows the additional global nuclear electricity generating capacity planned between 1975 and 2005 that was never built due to cancellations. Had the plans proceeded, the world would have seen an additional 11 GWe of nuclear power commissioned every year. Of the 165 cancelled plants,

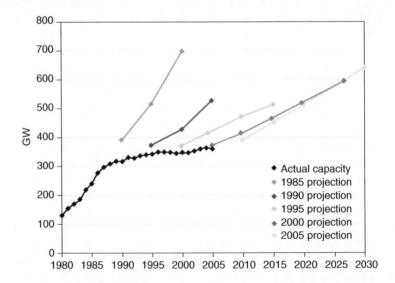

Figure 3.2 IAEA projections of world nuclear power capacity (high estimates) versus actual capacity (source: adapted from 'Projection of the world's nuclear power capacity in the high estimates', IAEA 2007b: 54).

construction had started on 62 and some were completed but never commissioned (IEA 2008a: 300).

In addition old plants are periodically closed as they reach the end of the planned life span. Some of the closures since the early 1990s have been for historic reasons that are unlikely to be repeated, notably the admission of former Soviet bloc states, Bulgaria, Slovenia and Lithuania, to the European Union, which required the shutdown of their old Soviet reactors. Even so, the world's nuclear fleet is old. By January 2008 there were 342 reactors aged 20 years or older (78 per cent of the total) (NEA 2008: 49). A major industrial effort will thus be required just to replace the current fleet – notwithstanding the possibility of life extensions to some existing plants of up to 30 years and maybe more.

A final compounding problem is that global estimates produced by bodies like the IAEA, the IEA and the NEA are used by others without reference to the caveats attached to them in their original form. The IEA, for example, in its 2008 report *Energy Technology Perspectives*, uses figures from the WNA for a chart on 'Plans and proposals for new nuclear power reactors' without mentioning how these are derived (IEA 2008a: 298–299). The WNA naturally has a vested interest in promoting the greatest accretion in nuclear energy possible. Its *World News Report* of September 2008 thus emphasized the IAEA's high rather than low projections in reporting that 'Nuclear Capacity Could Double by 2030' (WNN 2008c).

Predictions about the demand for nuclear power in the developing world are especially problematic. It is sobering to consider a special *Market Survey for Nuclear Power in Developing Countries* issued by the IAEA in September 1973. It concluded that 'the projected markets for nuclear plants which *will* [emphasis added] be commissioned' in 14 participating developing countries would total, by the end of the 1980s, 52,200 MW in the low estimate and 62,100 in the high (IAEA 1973: 5). Table 3.1 compares the predictions to the reality. The only

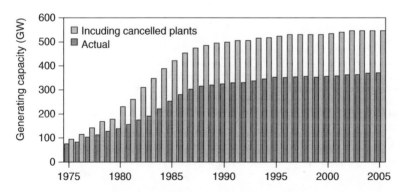

Figure 3.3 Planned global nuclear generating capacity versus actual capacity, 1975–2005 (source: IAEA, 2006, cited in IEA 2008a: 300).

states that ended up with nuclear power were the ones already engaged in build-ing a reactor – Argentina, Mexico, Pakistan, South Korea and Yugoslavia. Of the rest only one – the Philippines – started to build a plant and then stopped. Some of these states are, 30 years later, again talking about acquiring nuclear energy.

The revival so far

If one dates the revival of interest in nuclear energy from 2000, it is clear almost a decade later that progress has been slow. Several countries, notably in East Asia, have begun building new reactors as part of ambitious nuclear energy programmes, but many others have only announced intentions or plans, are studying the possibilities or are simply floating ideas. The first decade of the twenty-first century has seen the opposite of a revival in nuclear power. There has, in fact, been a relative decline, and, according to some indices, an actual decline, in the contribution of nuclear power to world energy produc-tion. Not only has the number of operating nuclear reactors plateaued since the late 1980s, but the IAEA figure of 436 reactors as of December 2009, with a total net installed capacity of 370 GW(e) (IAEA 2010c), is eight units less than the historical peak in 2002 of 444 (Schneider and Froggatt 2008: 4). Five nuclear power reactors remain in long-term shutdown. Since commercial nuclear energy began in the mid-1950s, 2008 was the first year that no new nuclear plant was connected to the grid (Schneider *et al.* 2009: 5), although two were connected in 2009.

Table 3.1 Predicted versus actual additions to nuclear generation capacity, 1980–1989

Country	Predicted capacity		Actual capacity	Number of reactors
	Low	High		
Argentina	6,000	6,000	600	1
Bangladesh	0	600	0	0
Chile	1,200	1,200	0	0
Egypt	4,200	4,200	0	0
Greece	4,200	4,200	0	0
Jamaica	0	300	0	0
Mexico	14,800	14,800	650	1
Pakistan	600	600	0	0
Philippines	3,800	3,800	0	0
Republic of Korea	8,800	8,800	6,977	7
Singapore	0	2,600	0	0
Thailand	2,600	2,600	0	0
Turkey	1,200	3,200	0	0
Yugoslavia	4,800	9,200	666	1
Total	52,200	62,100	8,893	10

Source: IAEA 1973.

In absolute terms, nuclear grew between 2000 and 2008 from 2,582.90 TWh to 2,738.60 TWh, a 6 per cent increase (BP 2009). This was dwarfed by a much greater growth in overall electricity generation during the same period, from 15,401.20 TWh in 2000 to 20,201.80 TWh in 2008, a 31.17 per cent increase. In the same period, nuclear's share of global generation fell from 13.5 per cent to 16.7 per cent, a decline of 3.21 per cent. Even this level was only sustained due to capacity factor improvements in the existing fleet and extended operating licences (mostly in the United States, where reactors set a generation record of 843 million gross MWh and averaged an historical high of 91 per cent capacity) (Ryan 2008: 1). *Nucleonics Week* described the causes as 'ranging from an earthquake in Japan to persistent aging ills in the UK to backfitting outages in Germany' (Ryan 2008: 1). But the decline has longer-term roots than that, as demonstrated by Figure 3.5 showing nuclear grid connections peaking in the 1980s.

Current 'new build'

The number of nuclear power plants currently being built appears impressive. According to the IAEA, 57 were 'under construction' worldwide in April 2010 (IAEA 2010c). Further analysis reveals, however, that current construction activity is confined to 15 countries, all of them, except Iran, with existing commercial nuclear power. Most of the activity – 40 reactors – is taking place in just four countries: China, India, Russia and South Korea. China is building almost half of these, so to that extent is leading the revival.

Almost one-third of what seems like new construction activity is in fact 'hang-over' orders from previous eras (Thomas *et al.* 2007: 11). Fourteen of the

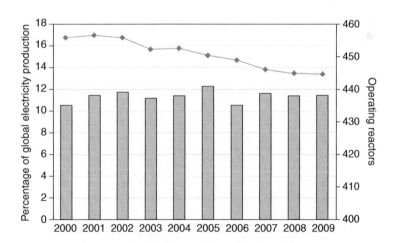

Figure 3.4 Nuclear reactor numbers and share of global electricity since 2000 (sources: IAEA 2010c; British Petroleum (BP) 2009).

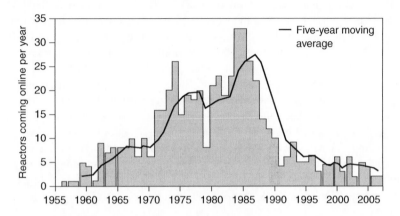

Figure 3.5 Global annual grid connections on five-year moving average (source: IAEA, 2006, cited in IEA 2008a: 301).

units 'under construction' have been designated by the IAEA as 'under construction' for 20 years or more (Schneider and Froggatt 2008: 8). Nine reactors currently on the 'under construction' list, in Argentina, Bulgaria, Russia, Slovakia and Ukraine, were on an IAEA list in December 2006 as 'nuclear power plants on which construction has been stopped'. The list does not include the Philippines as no decision has yet been taken to resume construction.

The two reactors in Bulgaria previously on the 'stopped' list, Belene-1 and Belene 2, will help replace four old Soviet RBMK reactors which Bulgaria shut down as a condition of EU membership (NEA 2008: 63). Two partially finished reactors in Ukraine, Khmelnitski-3 and Khmelnitski-4, which were, respectively, 75 per cent and 28 per cent complete when work stopped in 1990, were also on the 'construction stopped' list. As the Ukrainian government itself has announced that work will not actually recommence on them until an unspecified date in 2010, they should not yet be on the 'under construction' list (WNA 2009c). As for Russia, it is difficult to obtain information as to whether construction work is actually occurring on some reactors listed as 'under construction' or whether site maintenance is simply being carried out. The seemingly impressive total of nine for Russia is distorted by the inclusion of two small floating reactors of just 32 MW each (IAEA 2010c).[6] Iran is about to bring its Bushehr reactor online, but only after 35 years of periodically interrupted construction (Bahgat 2007: 20–22).[7] The resuscitation of previously defunct projects could be described as a revival of sorts, but is probably not what the nuclear industry has in mind; rather they are pinning their hopes on what they call 'new build'.

In terms of technology, most of the reactors currently under construction are based on old technology, in some cases slightly modified. The only Generation III reactors presently in operation are four ABWRs in Japan which went online in 1996, with two more under construction in Taiwan. No Generation III+ reac-

Table 3.2 Nuclear power plants under construction, April 2010

Country	Number of units	New nuclear capacity (MWe)	Existing units	Existing nuclear capacity (MWe)	% total capacity
Argentina	*1	692	2	935	3.5
Bulgaria	*2	1,906	2	1,906	19.6
China	23	23,620	11	8,438	1.9
Finland	1	1,600	4	2,696	16.0
France	1	1,600	59	63,260	54.4
India	4	2,506	19	4,189	2.2
Iran	*1	915	0	0	0.0
Japan	1	1,325	54	46,823	17.7
Pakistan	1	300	2	425	2.4
Russia	*9	6,894	31	21,743	10.5
Slovakia	*2	810	4	1,762	29.9
South Korea	6	6,520	20	17,705	24.1
Taiwan	*2	2,600	6	4,980	12.6
Ukraine	*2	1,900	15	13,107	25.7
United States	*1	1,165	104	100,747	10.1
Total	57	54,562			

Source: IAEA 2010c.

Note
* Denotes construction on previously suspended projects.

tors are currently operating, and only two types advertised as Generation III+ are under construction. Two EPRs are being built, one in Finland and one in France. The first AP1000 commenced construction in China in 2009 at Sanmen in Zheijiang province (WNN 2009b). No new CANDUs have commenced construction or even been ordered.

In future the market for new nuclear plants to 2030 will be dominated by large (1,000 MW and above) light water reactors with both Generation III and III+ characteristics (MIT 2003).[8] Areva's market research apparently indicates that about half the global demand is for large reactors between 1,350 and 1,700 MW and the other half is for what it calls 'midsize' reactors of 1,000 to 1,350 MW (MacLachlan 2009a: 5). However the number of new generation reactors built and their global spread will depend, at least in market economies, on fulfilling their promised advantages in reducing capital costs and construction times. As the 2009 MIT study asks of the United States:

> Will designs truly be standardized, or will site-specific changes defeat the effort to drive down the cost of producing multiple plants? Will the licensing process function without costly delays, or will the time to first power be extended, adding significant financial costs: Will construction proceed on schedule and without large cost overruns? … The risk premium will be eliminated only by demonstrated performance.
>
> (MIT 2009)

Plans for 'new build' by existing nuclear energy states

Plans for real 'new build' have been announced by 19 of the 31 countries that already have nuclear power, so to that extent a revival is occurring. Especially extensive are the plans of China, India, Japan, Russia, South Korea, Ukraine, the UK and the United States. However, close examination of each country's plans elicits caution.[9] Only China is both planning and undertaking the type of new build programme that could be described as 'crash', but it is starting from such a low base that it is not so much a renaissance as an initial foray into acquiring a substantial nuclear energy capability. India has never been able to fulfil its consistently expansive nuclear energy dreams. The United States could, in theory, mount a crash programme, as some of the nuclear industry's more ardent supporters advocate, but its federal/state division of power, deregulated market, mounting public debt, environmental regulations and still strong anti-nuclear movement are likely to slow it.

Canada

While Canada had plans for major new nuclear build, these were set back significantly in June 2009 with the announcement by the Ontario government that it was indefinitely postponing its decision on a new fleet of reactors (Cadham 2009).[10] Ontario has 16 of Canada's 18 nuclear reactors, supplying approximately 50 per cent of provincial electricity. The provinces of New Brunswick and Quebec have one each. In July 2009, Bruce Power, an electricity utility, also dropped plans to build new reactors in Ontario due to declining electricity demand (Bruce Power 2009). Meanwhile, the federal government announced in December 2009 that it would privatize the CANDU reactor supply part of Atomic Energy of Canada Ltd (AECL) (NRCan 2009), putting in question the future commercial prospects of the company's new Advanced CANDU Reactor (ACR).[11] Mooted new build in Alberta and Saskatchewan is a long way from realization, although refurbishment of reactors in Ontario, New Brunswick and Quebec will likely all be accomplished (Cadham 2009). The refurbishments in Ontario are already over budget and behind schedule (May 2010).

China

China has the most ambitious programme of any country. A director of its Nuclear Energy Agency, Zhou Xian, has said that 'the golden time for China's nuclear power development has come', with some projections as high as 72 GWe by 2020 (compared to 8.2 GWe today), requiring the construction of 60 reactors in 11 years (WNN 2009a). The last country to carry out such a rapid nuclear energy expansion was the United States in the 1970s (Bradsher 2009). China is certainly moving rapidly, but from a very low base, with only 1.5 per cent of total generating capacity currently provided by nuclear. In June 2008 it had only 11 reactors, compared with Canada's 18 and the United States' 104 (NEA 2008:

49). Even its most ambitious plans will see an increase to just 5 per cent of total electricity generation capacity by 2020.

As an indication of the scale of China's electricity programme of all types, its construction programme has accounted for 80 per cent of the world's new generating capacity in recent years (*The Economist* 2010b: 72). Capacity added in 2010 alone will exceed the installed total of Brazil, Italy and the UK and come close to that of Germany and France (*The Economist* 2010b: 72). China is therefore the energy exception, including in the nuclear area, that proves the rule and Chinese experience should not be extrapolated to other countries, especially India, which has similar grand plans.

Already there are concerns in China about finance, labour shortages and costs (Hibbs 2008: 1, 10), as well as safety. In October 2009 Prime Minister Wen Jiabao ordered a quintupling of the National Nuclear Safety Administration's staff by the end of 2010 to 1,000 (Bradsher 2009). The director of the organization, Li Ganjie, has warned that 'At the current stage, if we are not fully aware of the sector's over-rapid expansions, it will threaten construction quality and operation safety of nuclear power plants' (Bradsher 2009). In addition there are concerns about corruption. In August 2009 the Chinese government dismissed and detained the president of the China National Nuclear Corporation, Kang Rixin, on a $260 million corruption case involving allegations of bid-rigging in nuclear power plant construction (Bradsher 2009).

France

France is often cited as a shining example that others should emulate in the nuclear electricity field. It has the highest percentage of nuclear electricity, has had the most intensive nuclear building rate and has the most extensive recent experience of nuclear build. Driven by energy security concerns, France added 54 reactors between the late 1970s and early 1990s, employing a highly standardized design (NEA 2008: 320). Annual generation of nuclear electricity grew by 43 per cent over a 14-year period after 1990. Today nuclear provides 77 per cent of France's electricity, more than any other country. It is the second largest producer of nuclear electricity after the United States and the world's biggest exporter of it, mainly to Belgium, Italy and Germany (although in peak winter periods France is forced to import fossil-fuel generated electricity from the latter to cover its shortfall due to overuse of electric space heating) (Schneider 2008a).

Currently, however, France has only a modest domestic expansion plan, since its capacity is already substantial. Only one new reactor is being built in France, by EdF at Flamanville. Like the Areva project in Finland, it is a 1,600 MWe EPR intended to demonstrate the superiority of the Generation III+ reactor and like the Finnish project it is experiencing difficulties. It is estimated to be 20 per cent more expensive than expected when the order was placed in 2005 (€4 billion compared with the original €3.3 billion) (Reuters 2009). A second 'quasi replica' of Flamanville-3 is proposed for Penly in southern France by 2020 (Reuters 2009). According to EDF, the second EPR would be more expensive, as savings on construction costs due to the 'learning curve' from Flamanville-3 would be offset by potentially higher site-related costs and a tighter market for materials and equipment (*Nucleonics Week* 2008: 2).

Given its track record of building reactors, France could, in theory, mount a new build programme to supply its neighbours with even more nuclear electricity than it currently does. This may be politically unacceptable, however, since accepting the risks associated with producing nuclear energy in return for national energy security is quite a different proposition to doing so for export earnings.

Moreover, France may not be able to emulate its past success at home due to changed circumstances. Deregulation of the electricity market is putting pressure on prices, which will affect the old way of doing business. The real cost of France's massive nuclear construction programme has never been revealed, but the government may not be prepared indefinitely to write the type of blank cheque demanded by such a programme. Problems familiar to other nuclear energy states, including lack of personnel, material shortages, cost overruns and construction delays, may also affect France's own programme. The country is likely to face a shortage of skilled workers (Schneider 2008a). Some 40 per cent of EDF's operators and maintenance staff will retire by 2015.

France's role in the revival is more likely to be in selling French reactors and related nuclear services abroad. Areva, a French multinational corporation with a global reach, describes itself a 'the world leader in nuclear power and the only company to cover all industrial activities in this field' (Areva 2009). In addition to the design and construction of nuclear reactors and supply of products and services for nuclear power plant maintenance, upgrades and operations, it also engages in uranium ore exploration, mining, concentration, conversion and enrichment, nuclear fuel design and fabrication and back-end fuel cycle activities such as treatment and recycling of used fuel, cleanup of nuclear facilities and 'nuclear logistics'.[12] Areva claims it can capture about one-third – slightly more than 100,000 MW – of the nuclear generating capacity that it calculates could be built by 2030 (MacLachlan 2009a: 5).

India

According to M.V. Ramana, India has had a largely unique nuclear trajectory (Ramana 2009). Ever since India's independence, its political leadership and technological bureaucracy have been committed to a future in which nuclear power plays a big role. While all governments are prone to exaggerating their nuclear plans, India is one of the most egregious examples. It has consistently trumpeted huge increases in nuclear power generation, only to see its plans crumble. In 1970, India announced a ten-year plan calling for the construction of 2,700 MW of nuclear capacity by 1980, a level it reached only in 2,000 (Ramana 2009: 5). In the early 2000s, India's Department of Atomic Energy (DAE) projected 20 GW by 2020 and 275 GW by 2052, the latter amounting to 20 per cent of India's total projected electricity generation capacity.

None of these past plans has materialized, and the programme has been marred by accidents, poor safety practices, cost overruns and construction delays (Ramana 2009). As elsewhere, nuclear electricity has been expensive, a huge problem in a developing country with surging electricity demand and multiple demands on scarce capital. In part because of a shortage of domestic supplies of cheap and easily mined uranium, but also due to its attempts at industrial self-reliance, India is one of the few countries that envisages a nuclear expansion that is ultimately based on fast breeder reactors. With vast deposits of thorium, India has also for decades been researching ways of using it for commercial power generation, but so far without success.

India currently has 19 reactors (one in long-term shutdown, with another two closed for lengthy repairs), with a generation capacity of approximately 4,120 MW, just 2.2 per cent of the country's total electricity generation capacity (WNA 2010e). Four reactors with a combined capacity of 2,572 MWe are currently under construction. India's reactors have almost all operated vastly under capacity (all but two of them between 24 and 58 per cent) and uranium shortages due to the 30-year suppliers' embargo against India have led to shutdowns and delayed commissioning of reactors (Saraf 2008: 1, 4).

Following the September 2008 decision by the Nuclear Suppliers Group to exempt India from restrictions on most sales of civilian nuclear technology and materials, estimates of India's nuclear energy prospects have risen. Current plans are for 20 new reactors providing a total of 16,740 MWe (WNA 2010e). Several countries have signed nuclear cooperation agreements with India, including Canada, France, Russia and the United States, in the hope of supplying reactors. Russian firms are already building two reactors at Kudankulam in Tamil Nadu and four more are envisaged for that site, as well as two elsewhere. Sites are also being designated by the Indian government for additional reactors to be built by Areva, Westinghouse and Hitachi (*Nucleonics Week* 2009b: 1, 12). India is also relaxing its rules to allow new companies to build nuclear power plants. Presently only the government-controlled Nuclear Power Corporation of India Ltd (NPCIL) and Bhavini, a government enterprise set up to build fast breeder reactors, have been allowed to control nuclear infrastructure (WNN 2010b). Canada, France and Russia have reached agreements to sell uranium to India, although Australia has declined on non-proliferation grounds, despite having concurred in the NSG decision. India is currently building a 500 MWe Prototype Fast Breeder Reactor, intended to be the first of several to be constructed over the coming decades, although concerns have been expressed by Indian researchers about their safety and cost (Kumar and Ramana 2008: 87–114).

The Atomic Energy Commission chairman has promised that nuclear power will contribute 35 per cent of India's total electricity generation capacity by 2050. Since the DAE has projected that India will have an installed capacity of 1,300 GW (a nine-fold increase from the current 145 GW) by that time, the 35 per cent prediction implies that installed nuclear capacity would amount to

455 GW, more than 100 times today's figure (Ramana 2009: 5–6). Based on past experience, such an increase seems highly unlikely even given the sudden availability of foreign technology and assistance. As World Bank official Surya Sethi says, it will take the country a while to deal with the numerous challenges facing its power sector generally, including bureaucratic red tape, the widespread theft of electricity and politically motivated caps on electricity prices: 'You can't suddenly make India a China' (Bajaj 2010).

India is also planning to enter the export market. In the past India has constructed its own reactors based on the CANDU heavy water type which it has re-engineered and proposes to sell to others. In September 2009 it announced that it is developing for export a special 300 MWe Advanced Heavy Water Reactor (AHWR), adapted to use low enriched uranium instead of the mixture of uranium-233 and plutonium derived from thorium that it proposes to use domestically (WNN 2009h).

Japan

At the end of 2009 Japan had 53 commercial nuclear power plants in operation with total gross power generation capacity of 48.1 GW (NEA 2009c). In July 2008 the country adopted an Action Plan for Achieving a Low-Carbon Society that envisages that by 2020 over 50 per cent of Japan's electricity will be from 'zero emission sources', including nuclear power. However the specific plans for nuclear remain relatively modest, involving increasing the utilization capacity to match the best performing countries and the promotion of the 'steady construction of new facilities while ensuring complete safety as a fundamental premise' (NEA 2009c). Japan continues to stockpile plutonium and has the world's most advanced plans for operating a 'plutonium economy' in the quest for energy independence, but progress has been slow. Operation of its existing fleet has been plagued by earthquake-induced shutdowns, technical problems and scandals. As previously mentioned it has had difficulties introducing the use of MOX into its current reactors, as well as experiencing long delays in constructing its own fuel cycle facilities such as reprocessing plants and fast breeder reactors.

Russia

Russia has a large civilian nuclear fleet (31 units) and infrastructure that it inherited from the Soviet Union. In recent years it has been attempting to rationalize, restructure and modernize the industry with a view to building new reactors at home (currently nine are under various stages of construction, including two experimental floating units) and, perhaps more importantly, exporting them (see Chapter 1 for details). Russian nuclear engineering is robust and its designs simpler than the West's but it continues to suffer bad press as a result of the Chernobyl accident at a Soviet-designed plant, despite the fact that the accident was caused by operator error. Its behaviour towards its natural gas customers has also not provided reassurance to countries seeking energy security by buying Russian nuclear reactors.

Russia is thus facing significant challenges in realizing its nuclear ambitions. According to Miles Pomper,

> It is far from clear whether Russia will be able to fulfill its ambitious goals to more than double its electrical output from nuclear power, increase exports of nuclear reactors, and play an even larger role in providing fuel and fuel-related services for nuclear plants.
>
> (Pomper 2009: 2)

Russia's 15 RBMK Soviet reactors will need replacement despite extensive safety improvements since the Chernobyl disaster (NEA 2008: 450)[13] and attempted lifetime extensions (Pomper 2009: 4). Although Russia is planning to export nuclear reactors of various sizes, including novel types like small floating ones, and has already bid for new build in Turkey, as well as vying for additional sales in India and new build in Vietnam, it is not clear at this stage whether the Russians will be able to compete in deregulated markets or in the OECD countries generally where reliance on Russia for any form of technology, much less one as sensitive as nuclear reactors, may be anathema.

South Korea

Like China, the Republic of Korea is one of the more likely candidates to achieve its nuclear energy plans. Currently, its 20 nuclear reactors supply 40 per cent of the country's electricity; it is building six more and envisages bringing as many as 18 new units online by 2030 and 60 by 2050. It also seems determined to establish a closed nuclear fuel cycle on the basis of spent fuel pyroprocessing (currently being perfected by South Korean scientists, but arousing US concerns about its potential proliferation implications) and fast reactors (Nuclear Fuel 2008: 5).

After being reliant on importing nuclear reactor technology (Westinghouse and CANDU) since the 1970s, the South Koreans since 2008 have been able to design, manufacture and build two types of 1,000 MW pressurized water reactors: the 1,000 MW Optimised Power Reactor (OPR) 1000 and the 1,400 MW Advanced Power Reactor (APR) 1400. There are also plans for an APR+ by 2012, as well as small- and medium-sized reactors for export (Mee-young 2010). Four OPR-1000s and two APR1400s are currently under construction, with an additional two APR1400 units to be built beginning in 2015 and 2016 (WNA 2010c). Korean reactor builders have a reputation for being able to adhere to construction timetables or even come in under schedule. The founder of Korea's nuclear programme cites an average construction time of 53 months, which is relatively fast by international standards.

South Korean firms, supported by the government, are mounting aggressive sales campaigns to export their nuclear reactors. Indonesia, Turkey and Vietnam are among those wooed as potential customers. South Korea announced its first reactor sales, to the UAE, in December 2009 (WNN 2009i). In addition to the Korean Electric Power Company (KEPCO), the winning consortium included Samsung, Hyundai and Doosan Heavy Industries, all South Korean firms, as well as Westinghouse and Toshiba. The Korean government has a 21 per cent share in KEPCO, while the newly privatized Korea Finance Corporation (KOFC) has a 30 per cent share (WNA 2010c). This is the first time a developing country has exported a nuclear power plant and may start a trend that overturns assumptions about the barriers to entry for new suppliers, especially if the quoted low prices can be maintained, quality assured and planned construction times achieved.

Ukraine

Ukraine might be considered as having an especially urgent motivation for nuclear 'new build' in its desire to escape reliance on Russian gas supplies which have been disrupted in recent years. Currently, Ukraine has 15 reactors, all of the Soviet-designed VVER type, generating about half its electricity. It plans to build 11 new reactors and nine replacement units to more than double its nuclear generating capacity by 2030. However, the drawn-out financing and safety enhancement saga over resuming work on Khmelnitsiki 3 and 4, involving the European Bank for Reconstruction and Development (EBRD) and Russia (WNA 2009c), as well as Ukraine's parlous economic situation, do not inspire confidence that this schedule is achievable. Even the construction of a new sarcophagus over the damaged Chernobyl reactor has been plagued by the instability of the Ukrainian government, incompetence and corruption, notwithstanding the willingness of foreign donors, such as Canada, to provide millions of dollars towards the project.

United Kingdom

The UK currently has 19 reactors, which generated 15 per cent of its electricity in 2007, down from 25 per cent in recent years due to plant closures. The former Labour government's January 2008 Energy White Paper, after an extensive public consultation process, concluded that nuclear power could be part of a low-carbon energy mix needed to meet the country's carbon emission targets. However, the government was careful to stress that it would be 'for energy companies to fund, develop and build new nuclear power stations in the UK, including meeting the full costs of decommissioning and their full share of waste management costs' (Hutton 2008). The plans call for commencing construction of the first new British nuclear power station in decades in 2013–2014, for completion by 2018. Ian Davis notes that 'previous British experience with untried nuclear designs suggests it could be much longer' (Davis 2009: 30). Crucially, in the UK's deregulated market, investment decisions are largely being left to private sector energy companies. The UK regulator, meanwhile, has expressed safety concerns about both prime contenders for the UK's new reactors, the EPR and Westinghouse's AP-1000. EDF, a front-runner for the UK's new build stakes, is purchasing land, mostly near existing plants, for its intended nuclear power fleet. The replacement of the Labour government in May 2010 by a coalition Conservative–Liberal Democrat one brings new uncertainty to new build in the UK since the Liberal Democrats have traditionally opposed nuclear power (some elements quite vehemently), while the Conservatives have supported it.

United States

The United States has the largest civilian nuclear fleet in the world, its 104 units providing approximately 20 per cent of the nation's electricity. The home of the original nuclear energy boom in the 1970s and 1980s, it has not built a new reactor in 30 years. Nonetheless, due to the size of its electricity market and influence in nuclear matters generally, it has been seen as a bellwether of the purported nuclear 'renaissance' following the Bush administration's launch of its Nuclear Power 2010 programme in 2002.

Since then, however, the United States has added only one new reactor to the grid, a shutdown plant at Browns Ferry that was refurbished and restarted. A reactor that was previously ordered but never completed, Watts Bar 2, is currently being finished (MIT 2009: 4–5). No new reactors are under construction. As of September 2009, 17 applications for licences to construct and operate 26 new reactors had been filed with the Nuclear Regulatory Commission (NRC), but even industry promoters predict that only four to eight new reactors might come online by 2015, and then only if they secure government loan guarantees (NRC 2009b). The Congressional Budget Office concludes that it is probable that 'at least a few nuclear power plants will be built over the next decade', most likely in states where electricity usage and the corresponding demand for additional baseload capacity are expected to grow significantly (CBO 2008: 26).

The 2003 MIT study argued that for nuclear power to be resurgent in the United States 'a key need was to design, build and operate a few first-of-a-kind nuclear plants with government assistance, to demonstrate to the public, political leaders and investors the technical performance, cost and environmental accountability of the technology' (MIT 2009: 19). The George W. Bush administration had offered several incentives to the nuclear industry, including streamlining the approval process for new build (with combined construction and operational licensing) and loan guarantees. The 2009 update of the MIT report lamented, however, that governmental support had been insufficient and ineffective (MIT 2009: 18).

The Obama administration has moved to provide some increased support to the nuclear industry by tripling the amount provided for loan guarantees by adding $36 billion to its 2011 fiscal year budget proposal (Spotts 2010). The first loan guarantee of $8.3 billion has been provided to the Atlanta-based Southern Company for two reactors at its Vogtle generating station near Augusta, Georgia that will cost $14 billion (originally priced at $1 billion for four reactors) (Spotts 2010). This is dependent on the NRC granting a licence to build and operate the plants. Even with loan guarantees local electricity bills are expected to increase by 9 per cent to pay for the new reactors. The state government, meanwhile, has passed a law ensuring that ratepayers will not be repaid if the utility fails to complete the plant. All this is reminiscent of the problems that plagued the original nuclear energy surge in the United States.

For many analysts even the expanded loan guarantee programme will still not be sufficient to kick-start the US nuclear industry since such guarantees are predicated on investors actually deciding to invest. Direct subsidies are not on the horizon. Peter Bradford, a former member of the Nuclear Regulatory Commission, has calculated that of the 26 new applications submitted to the NRC since 2007, nine have been cancelled or suspended indefinitely, and ten more have been delayed by one to five years (Grunwald 2010). Utilities like Exelon, Duke Energy and Florida Power & Light have cancelled· or scaled back their nuclear power proposals.

The main deterrent is that prices are rising dramatically, causing 'sticker shock' according to NUKEM, Inc., a company that tracks 'The people, issues and events that move the fuel market'. It reported in April 2008 that with projects now 'spiralling upwards to a dizzying $7 billion per reactor with all-in costs in the range of $5,000 to $7,000/kWe', the ' "early" nuclear renaissance in America now looks more like 2015–2020 instead of our originally designated 2013–2017 period' (NUKEM Inc. 2008: 2–4). *Time* magazine reported that in 2009 estimates for several reactors doubled, while the price of one in Pennsylvania more than tripled (Grunwald 2010). As Sharon Squassoni concludes:

> just to maintain its share of the electricity market, the nuclear industry would need to build 50 reactors in the next 20 years. Given that only four new reactors might be operational by 2015, significant growth could require build rates of more than four per year. Greater government subsidies and a carbon pricing mechanism are not likely enough to achieve such rates of construction. The best outcome for the US nuclear industry over the next five years, particularly under an administration that will probably offer mild rather than aggressive support, will be to demonstrate that it can manage each stage of the licensing, construction and operating processes of the first reactors competently and efficiently. In sum, the industry needs to demonstrate that it has overcome the problems of the past.
>
> (Squassoni 2009b: 18)

Other current players

Other states with existing nuclear plants have also announced new build plans, but have not yet begun to implement them. They include Brazil, Romania and Lithuania (in partnership with Estonia, Latvia and Poland). Slovakia is currently building two reactors. South Africa has recently cancelled its expansion plans due to its financial situation (*Nucleonics* Week 2008: 1).[14] The states with existing nuclear power that are not currently planning new build include Belgium, the Czech Republic, Germany, Hungary, Mexico, the Netherlands, Slovenia and Spain.

The situation in Europe is particularly mixed. Only 15 of the 27 EU member states currently operate nuclear power reactors, although nuclear power provides one-third of the EU's total electricity. Several EU states, notably Austria, Ireland and Luxembourg, oppose nuclear energy, which complicates the addition of new build in Europe. Currently, of the European states that decided to phase out nuclear power after Chernobyl – Belgium, Germany, Italy, Spain, Switzerland and Sweden – only, Italy, Germany and Sweden have reversed or partially reversed their positions.

Italy is planning a whole new fleet of reactors after closing its entire original fleet after Chernobyl. In Sweden, the Conservative-led coalition government decided to halt the phase-out plans for Sweden's ten ageing nuclear reactors and to replace decommissioned nuclear plants with new ones (although not adding additional units). The opposition, however, has reiterated its collective support for the 1980 referendum that led to the current gradual phase-out of nuclear power in the country (*Nucleonics Week* 2009c: 8; Bergenäs 2009). Sweden's future policy thus remains uncertain. Upgrading and refurbishment of its existing old reactors has not gone well, leading to record high electricity prices in 2009–2010 (EurActiv.com 2010). In Switzerland companies are applying to replace several first-generation units with new, larger, advanced plants (WNA 2010d). However public opinion is divided over nuclear energy and the approval process at both the cantonal and federal levels may be protracted. Germany may now extend the lifetimes of its existing reactors.

Aspiring nuclear energy states

The WNA suggests 'over thirty countries' are newly interested in nuclear energy (WNA 2010b). The IAEA has publicly claimed that 'A total of 60 countries are now considering nuclear power as part of their future energy mix, while 20 of them might have a nuclear power programme in place by 2030' (IAEA 2009e). The number 60 bears an uncanny resemblance to the total number of states that have recently approached the IAEA, at any level and in whatever detail, to discuss nuclear energy. The IAEA's Director-General of Nuclear Energy, Yury Sokolev, told representatives of 40 countries attending an agency workshop on IAEA Tools for Nuclear Energy System Assessment (NESA) for Long-Term Planning and Development in July 2009 that the IAEA is expecting to assist 38 national and six regional nuclear programmes, a 'three-fold increase from the previous [unidentified] reported period' (IAEA 2009e). This, of course, does not mean that so many states will decide to proceed with nuclear energy after conducting their assessments. The number may also include states that already have nuclear power. Given the IAEA's caution that the timeframe from an initial state policy decision (a nebulous concept itself), to the operation of the first nuclear power plant 'may well be 10–15 years' (IAEA 2007c: 2), a sudden surge in nuclear energy capacity in the developing world by 2030 seems inherently unlikely.

The Survey of Emerging Nuclear Energy States (SENES),[15] compiled by the Nuclear Futures Project of the Centre for International Governance Innovation (CIGI) in Waterloo, Ontario, and the Canadian Centre for Treaty Compliance (CCTC) at Carleton University in Ottawa, tracks progress made by states that have declared a serious interest in acquiring a nuclear energy capability – from the first official announcement of such an interest to the connection of the first nuclear power plant to the country's electricity grid. SENES reveals that just 34 states, plus the members of the Gulf Cooperation Council (GCC) collectively, have announced 'consideration' or 'reconsideration' of nuclear energy at a credible ministerial level since 2000. SENES lists fewer states than are identified in other surveys, reflecting a more sceptical view of state intentions than is likely to be adopted by a multilateral organization or an industry association.

In all of the surveys of states allegedly interested in nuclear energy, the vast majority are developing countries. In the case of SENES only four (Israel, Italy, Poland and Turkey) could be considered developed. A couple of others (Belarus and Malaysia) could be considered developed enough to be able to afford a nuclear power plant, although whether they have the other prerequisites is doubtful. Several states could be considered independently wealthy enough as result of oil income to be able to afford a nuclear reactor on a turnkey basis: these include Algeria, Indonesia, Libya, Nigeria, Venezuela and the Gulf States, including Saudi Arabia and the UAE. But all of them lack an indigenous capacity at present to even operate, regulate and maintain a single nuclear reactor, much less construct one.

To track states' progress, SENES uses some of the key steps set out in the IAEA's *Milestones in the Development of a National Infrastructure for Nuclear Power*. This document identifies three broad categories of achievements which must be accomplished before a state is considered ready for a nuclear power programme (IAEA 2007c):

Milestone 1 – Ready to make a knowledgeable commitment to a nuclear program.
Milestone 2 – Ready to invite bids for the first nuclear power plant.
Milestone 3 – Ready to commission and operate the first nuclear power plant.

The vast majority of the states identified in SENES could not, at present, legitimately claim to have reached or gone beyond Milestone 1. Only Iran is close to starting up a reactor and connecting it to the grid (probably in late 2010). Save for this one exception, none has begun construction. The Philippines has a partially completed reactor in Bataan, which it may resume work on. Of the rest only Italy, which was among the pioneers of nuclear technology and had a nuclear power industry before scrapping it after the Chernobyl accident, could be said to be completely knowledgeable about nuclear power requirements. As

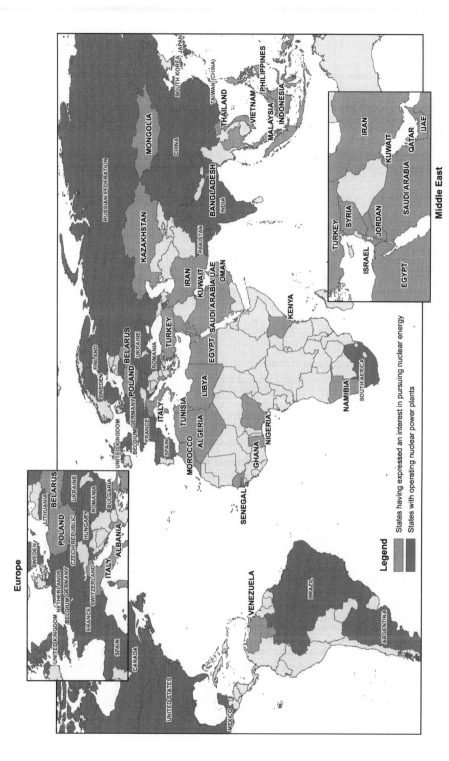

Map 3.1 Current and aspirant nuclear energy states (source: *Survey of Emerging Nuclear States (SENES)*, Centre for International Governance Innovation, http://cigionline.org/senes. IAEA 2010c)

of May 2010 only Egypt, which has aspired to nuclear power for more than 30 years, Turkey and the UAE are known to have invited bids for a plant, which puts them at Milestone 2.

Many of the aspiring states listed in SENES have taken some steps, such as consulting the IAEA, conducting a feasibility study or establishing an atomic energy commission and/or nuclear regulatory authority, the latter generically known by the IAEA as a Nuclear Energy Programme Implementing Organization (NEPIO). But this is less impressive than it may appear, as these are among the easiest steps and imply nothing about the capacities of such countries to actually mount a nuclear energy programme.

Getting beyond Milestone 1 poses increasingly more difficult challenges. While many of these may be difficult to quantify, some indicators can almost certainly rule out countries from being capable of acquiring nuclear power over the course of at least the next two decades. Such countries would need to make unprecedented progress in their economic development, infrastructure and governance before nuclear power is a feasible option. The unpreparedness of most SENES countries is revealed by three measurable indicators: governance, physical infrastructure and finance.

Governance

Many SENES states struggle with governance generally. Shockingly, all SENES developing states except Qatar, the UAE and Oman, score five or below on the ten-point scale of Transparency International's Corruption Perception Index (Transparency International 2009). Considering that the institutional framework for a successful nuclear energy programme critically includes an independent nuclear regulator, free from political, commercial or other influence that must ensure the highest standards of safety and security, the implications of pervasive corruption in potential new entrant states are frightening. World Bank indicators for political violence, government effectiveness, regulatory quality and control of corruption (see Figure 3.7) show a high correlation between these factors. Negative indicators of regulatory control characterize almost all of the SENES states. Nigeria, for example, has a long history of mismanaging large, complex projects (Lowbeer-Lewis 2010: 18), partly due to governance deficits, but also corruption, so establishing the regulatory infrastructure and safety culture for a nuclear power plant even over a ten-year period poses an immense challenge.

As figures from the World Bank for 2008 illustrate, SENES states generally fare poorly in terms of political violence, government effectiveness, regulatory control and control of corruption. Regulatory control is especially important in ensuring nuclear safety, security and the proper implementation of nuclear safeguards. At least half the SENES states are in the negative range for most of the indicators in Figure 3.7.

A tiny number of states can probably successfully purchase everything necessary for a nuclear power programme, including safety and security personnel for their institutional infrastructure. All of the Gulf States, like the UAE, Qatar and

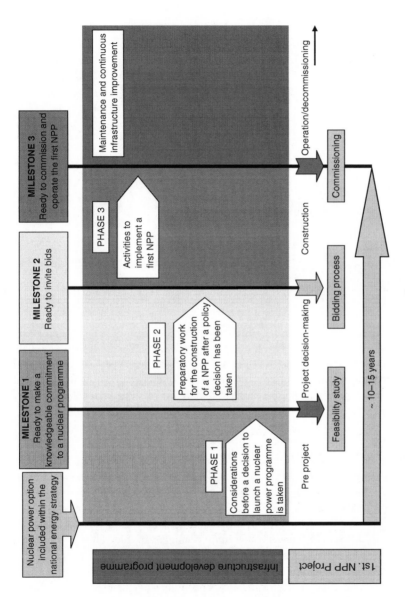

Figure 3.6 Infrastructure development programme (source: IAEA 2008f: 5).

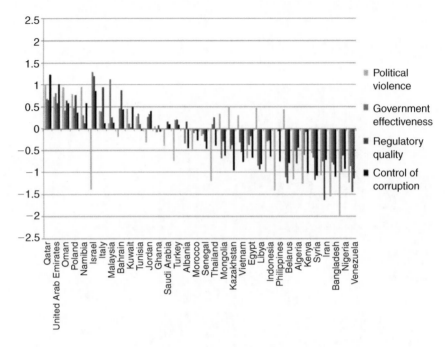

Figure 3.7 Governance indicators for SENES States, 2008 (source: World Bank 2009).

Saudi Arabia, are used to using contractors throughout their economy and may do so in the nuclear case. Not many other SENES states can afford to emulate them. Yet even these countries cannot purchase the types of institutional structures and governance norms and practices developed over many years that are required for managing complex undertakings like a nuclear energy programme.

Physical infrastructure

As discussed in Chapter 2, a major barrier to aspiring nuclear states in the developing world is having the physical infrastructure to support a nuclear power plant or plants. This includes an adequate electrical grid, roads, a transportation system and a safe and secure site. While it was impossible to survey all of the infrastructure capacities of the SENES states for this study, a single variable, the size of a state's installed generating capacity, serves as a useful proxy for its readiness to host a large nuclear power plant. The IAEA recommends that no more than 10 per cent of a country's electricity grid should be dependent on a single nuclear plant. Hence for a 1,000 MW plant or above (currently the size of Generation III and Generation III+ plants), a 9,000 MW capacity is recommended.

Based on installed electricity generation capacity for 2005[16] (see Figure 3.8) only 16 of the 34 SENES states currently have such a capacity. These are either

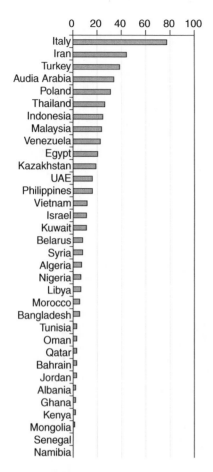

Figure 3.8 Generation capacity (GWe) (source: US Energy Information Administration 2009).

developed countries like Italy or large developing states like Indonesia and Kazakhstan. It is no coincidence that the three countries at the top of the table – Italy, Iran and Turkey – are the only SENES states which have clearly passed Milestone 1 and are among the most likely aspirants to succeed. To overcome such deficits the aspirant countries would need to spend millions of dollars on upgrading their existing electricity grids in addition to the expense of a new nuclear plant itself.

Three of these, Belarus, Syria and Algeria, are close to the cut-off point of 9,000 and may be able to increase their capacity while simultaneously planning a new power reactor. The remaining states – those with less than 6,000 MW capacity – would need to increase it by more than 50 per cent to bring it to the minimum 9,000 MW. Buying smaller size Generation III reactors is not currently an option as they are not yet technically proven, much less commercially available.

Linking national grids to others' or selling electricity are options for some states. Egypt, for instance, even though it has a sufficiently large national installed capacity, envisages selling electricity to its neighbours, and may in future be connected to the European grid (Shakir 2008). Those with less than 10,000 MWe, like Algeria, Libya, Morocco and Syria, may also eventually be linked to a European/Mediterranean grid, which would permit them to move ahead with their nuclear plans. Yet this adds a further layer of uncertainty to their aspirations. Jordan is already connected to a regional grid, which means its plans for a reactor, despite having a small national grid capacity, make some sense. Most of the African aspirants have small, poorly maintained and unconnected grids.

The Gulf Cooperation Council (GCC) comprising Bahrain, Kuwait, Oman, Qatar, Saudi Arabia and the UAE, all of which are SENES states, is pursuing a jointly owned nuclear plant that would supply electricity to all of the partners. Alone, many of them could not effectively and efficiently use a nuclear power plant due to their limited national power requirements. Bahrain, for instance, which says it is interested in its own nuclear power plant, has a total installed capacity of only 3,000 MW for a population of less than one million. While collectively they could use their oil wealth to purchase one or more reactors, there are doubts about the seriousness of their proposal. In addition, their grids are currently not well connected. The group has in fact been advised by the IAEA to make unifying investments in their electricity grids in parallel with any investment in a common nuclear reactor (Hibbs 2009: 8).

The Baltic States – Estonia, Latvia and Lithuania – and Poland were considering jointly building two 1,600 MW reactors to supply electricity to all four countries (*Nuclear Energy Daily* 2007). Even combined, the three Baltic States do not have enough generating capacity, so new nuclear was only considered viable as a joint project with Poland. The project has, however, at the time of writing, been grounded, largely because of political disagreements between Lithuania (the host country) and Poland (*Nucleonics Week* 2007a: 5). Similar problems could arise in other projects involving joint owners. Such joint ventures invariably add further complexity to already complex nuclear energy plans and projects.

Finance

Another main indicator that a state may not be able to follow through with its nuclear plans is its ability to finance a nuclear plant. As indicated in Chapter 2 a country's GDP and credit ratings are crude indicators of 'affordability'. States with both a low GDP and a poor credit rating are unlikely to be able to secure a loan for nuclear energy development. Table 3.3 displays the GDP and credit ratings for the 34 aspiring nuclear states listed in SENES. Only Israel, Italy, Poland and the Gulf States had A ratings. Nine states – Albania, Ghana,

Table 3.3 SENES states' GDP and credit ratings

State	2007 GDP (billion USD)	Credit rating
Italy	1,834.00	A+
Turkey	893.10	BB–
Indonesia	863.10	BB–
Iran	790.60	
Poland	636.90	A–
Saudi Arabia	553.50	AA–
Thailand	533.70	BBB+
Egypt	414.10	BB+
Malaysia	367.80	
Venezuela	357.90	BB–
Nigeria	318.70	BB–
Philippines	306.50	BB–
Algeria	228.60	
Vietnam	227.70	BB
Bangladesh	213.60	
Israel	197.50	AA
Kazakhstan	171.70	BBB–
United Arab Emirates	171.40	AA
Kuwait	137.40	AA–
Morocco	129.70	BB+
Belarus	104.50	B+
Syria	90.99	
Libya	83.59	
Tunisia	78.21	BBB
Qatar	76.75	AA–
Oman	62.97	A
Ghana	32.02	B+
Jordan	29.07	BB
Bahrain	25.17	A
Senegal	20.92	B+
Albania	20.57	
Namibia	10.87	
Mongolia	8.70	

Sources: GDP figures from Central Intelligence Agency 2009; credit ratings from Standard and Poor's 2009.

Jordan, Libya, Mongolia, Namibia, Senegal, Syria and Tunisia – had a GDP less than $100 billion in 2007, along with non-existent or uncertain credit ratings (BBB or lower). The possibility that a single nuclear reactor could cost up to $10 billion, more than one-tenth of each of these states' GDPs, illustrates the problem. It is no coincidence that the states identified as having insufficient grid capacity tend to be the same ones with a low GDP and non-existent or poor credit ratings.

The only developing countries that may be able to ignore such constraints are, again, those with oil-based wealth, such as Nigeria, Saudi Arabia, the Gulf States and Venezuela. Some may be able to afford to buy reactors outright without loans. Others, like the Gulf States, have, on the whole, good credit ratings and would be able to secure commercial loans. The recent drop in the price of oil and international financial turmoil are likely to make even these states wary of committing to expensive new infrastructure projects like a nuclear power reactor. The richest emirate in the UAE, Dubai, is reportedly $80–120 billion in debt, and has had four of its banks downgraded by Standard and Poor's credit rating agency (*The Economist* 2009d: 45).

The most likely new entrants

An accurate assessment of a country's probability of success in acquiring a nuclear power programme requires in-depth knowledge of the internal political dynamics of each individual state, especially how it makes its energy decisions on and finances energy projects. It can probably be assumed that new entrants among the developed states, Israel, Italy and Poland, will have little difficulty in acquiring nuclear energy from a technical, financial and institutional point of view. For them the greatest barriers could very well be political. The following analysis thus focuses on the prospects of the most likely candidates among the developing states. Among these Jordan would appear to be ruled out based on the criteria discussed above, but it does appear to be particularly determined and therefore warrants closer attention.

Algeria

In a statement by Energy Minister Chakib Khelil, Algeria announced its intention in 2008 to build a nuclear power plant within ten years (Moj News Agency 2008). To expedite this it has signed nuclear cooperation agreements with China, France, Russia and the United States (WNA 2010b). As Africa's largest natural gas producer, the country is looking to diversify its energy sources, including electricity generation from wind, solar and nuclear (GulfNews.com 2008), as well as for water desalination (Merabet 2009). The government established the Commissariat pour l'énergie atomique (Comena) in 1996 to cover a range of possible applications for nuclear energy, including in the agriculture and health sectors. Algeria also has an extensive nuclear research establishment, including two research reactors, a pilot fuel fabrication plant and various facilities at the Ain Oussera 'site', comprising an isotope production plant, hot-cell laboratories and waste storage tanks (IISS 2008: 107–113). Algeria was involved in controversy in the early 1990s over its nuclear weapons potential, especially as it was not an NPT party and its facilities were not safeguarded. Algeria ratified the NPT in 1995 and signed a full-scope safeguards agreement in 1997. Despite its initial steps and research capacities, Algerian nuclear energy plans are still in their infancy, and the country's 2018 target for a nuclear power plant is unrealistic, not least because its current electrical grid capacity of 6,470 MW is not nearly sufficient to support a nuclear power plant.

Egypt

With a longstanding interest in nuclear energy, Egypt has managed since the 1950s to establish four research facilities, including two research reactors, a fuel-manufacturing plant and a pilot conversion plant. However, after the Chernobyl disaster in 1986, it put its plans for a nuclear power programme on hold (Windsor and Kessler 2007: 13). It has since reinvigorated its efforts with President Hosni Mubarak's announcement in October 2006 that the country would once again try for a nuclear reactor to meet its energy needs. Several feasibility studies were conducted, leading to the announcement in January 2008 that a 1,000 MWe reactor would be built at El-Dabaa on the Mediterranean coast (*The Economist* 2007; Reuters 2007; WNA 2010b). Egypt has since taken several concrete steps, including preparing a site at El-Dabaa and putting out a call for bids for the plant's construction (*Egypt News* 2008). Once a bid is accepted, the Egyptians estimate the project will take ten years to complete at a cost of between $1.5 and $1.8 billion (*Global Security Newswire* 2008a). In May 2008 the government began assessing construction tenders (*Egypt News* 2008). So far no bid has been selected, no doubt because Egypt's price range is orders of magnitude lower than the $5–6 billion it typically costs to build a 1,000 MWe nuclear plant. Financial challenges loom large over Egyptian prospects for a nuclear power plant, and despite the country's recently improved credit rating, the possibility of attracting foreign investors to a major nuclear power project remains remote (IISS 2008: 28). As the International Institute of Strategic Studies notes, 'The Egyptian civil nuclear programme has often been described as "budding", meaning that it is both underdeveloped and under development at the same time' (IISS 2008: 24).

Indonesia

Almost since independence, Indonesia has sought to acquire nuclear energy, sometimes envisaging up to 30 reactors spread across its sprawling archipelago, but its lack of financial, organizational and technical resources has always held it back. Currently the National Atomic Energy Agency (BATAN) operates three research reactors. The newest, a 30 MW (thermal) unit at the Serpong Nuclear Facility near Jakarta started up in 1987, was intended to support the introduction of nuclear power to the country (WNA 2010b). Newly democratic and developing economically, Indonesia has revived its interest in nuclear energy. President Susilo Bambang announced in 2006 a decision to pursue a nuclear energy programme to meet rising energy demand (*Jakarta Post* 2007). Present plans are to build a single plant comprising two 1,000 MWe reactors on the Muria Peninsula in Central Java by 2017, with possibly two more later, bringing the total contribution of nuclear to 2 per cent of Indonesia's capacity (WNA 2010b). In July 2007 Korea Electric Power Corporation and Korea Hydro & Nuclear Power Company signed a memorandum of understanding with Indonesia's PT Medco Energi Internasional for a feasibility study on building two 1,000 MWe

Optimized Power Reactor (OPR)-1,000 units at a cost of $3 billion. Plans were to call tenders in 2008 for Muria 1 and 2, leading to decision in 2010, with construction starting soon after and commercial operation from 2016 and 2017, but this schedule has slipped by at least two years. The delay may be partly due to Indonesia's high levels of seismic activity, which not only requires careful site planning but has led to significant public opposition to the construction of the plant (Harisumarto 2007). Since the advent of democracy in the country a vibrant anti-nuclear movement has also been able to make its voice heard without fear of repression. The main problem, however, is that the central government has not yet formally approved plans for nuclear energy due to political and electoral politics (*Nucleonics Week* 2007b). An IAEA official stated in 2007: 'We don't see Indonesia moving this program forward' (*Nucleonics Week* 2007b).

Jordan

In a statement by King Abdullah II, Jordan announced in January 2007 that it would pursue a nuclear energy programme. With 2 per cent (112,000 tons) of the world's reasonably assured low-cost uranium supplies (WNA 2008b) Jordan is intending to begin mining operations, even though it is still unclear whether it intends to buy reactors fuelled by natural uranium or LEU. Jordan's Committee for Nuclear Strategy, set up in 2007, has set out a programme for nuclear power to provide 30 per cent of electricity by 2030 or 2040, and to provide for exports. Also established in 2007 were the Jordan Atomic Energy Commission (JAEC) and the Jordan Nuclear Regulatory Commission. In April 2007 the government entered discussions with the IAEA to assess the feasibility of a nuclear power plant (Reuters 2007). Since then the government has signed nuclear cooperation agreements with Canada, China, France, Russia, South Korea, the UK and the United States (WNA 2010b).

In May 2010 three vendors and designs were short listed, the Atmea-1 from Areva-Mitsubishi Heavy Industries, the AECL's Enhanced CANDU6, and the AES-92 from Russia's Atomstroyexport (WNA 2010b). The firm Worley Parsons will assist the JAEC with technology selection, tender preparation and evaluation of the bidders, as well as assisting in fuel cycle engineering and waste management for the plant. It will also assist in establishing a utility company, expected to be a public–private entity, to own and operate the plant. The JAEC expects to start building a 750–1,100 MWe nuclear power plant in 2013 for operation by 2020. A second is envisaged for operation by 2025, while in the longer-term, four more are envisaged. In December 2009 the JAEC selected a consortium headed by the Korean Atomic Energy Research Institute (KAERI) with Daewoo to build the country's first research reactor, a 5 MW unit at the Jordan University for Science and Technology by 2014.

Jordan's desire for nuclear energy is partially a result of its dependence on imports for 95 per cent of its current energy needs (Dow Jones Newswires 2009). With its small 2,098 MW electrical grid, low GDP and poor credit rating, the idea of Jordan successfully acquiring nuclear power seems implausible, except that it is planning to export electricity regionally, including to Israel and Egypt. The plant may be dedicated, at least in part, to water desalination, in which case the existing grid capacity is not as significant an issue. It is unclear, however, where the financing will come from, although Jordan has special characteristics that may help it obtain favourable loan terms and/or foreign aid. Jordan's close relationship with the United States, its friendly relationship with all its neigh-bours (unique in the Middle East) and its international reputation for moderation and diplomatic savvy may help it succeed in its nuclear energy plans where others fail – and may be the exception that proves the rule.

Turkey

Turkey is another country that has sought nuclear energy since the late 1960s, but its plans have always been stymied by financial considerations. Turkey has two research reactors, a small-scale pilot facility for uranium purification, con-version and production of fuel pellets, and a nuclear waste storage facility for low-level nuclear waste (IISS 2008: 63–64). It has negotiated nuclear coopera-tion agreements with several countries, both regionally and more broadly. A key agreement with the United States was approved by the US Congress only in June 2008 after being delayed because of concerns about Turkish companies' involve-ment with the A.Q. Khan nuclear smuggling network. In 2006 Turkey announced it was planning to build several reactors to produce 5,000 MW of electricity by 2015 (IISS 2008: 65). Construction was originally scheduled to begin on the first reactor in 2007. However, legal and tendering difficulties have led to continuing delays (*Nucleonics Week* 2009b: 6).

In May 2010 Russia and Turkey signed an intergovernmental agreement for Rosatom to build, own and operate four 1,200 MWe units at Akkuyu on the Mediterranean coast near the port of Mersin (WNA 2010b). Rosatom will finance the project and start off with 100 per cent equity, although the Turkish government has announced it will take a 25 per cent share in any nuclear power projects in Turkey. The state electricity company TETAS will buy a fixed pro-portion of the power at a fixed price of $12.35 cents/kWh for 15 years, with the remainder sold by the project company on the open market. After 15 years, when the plant is expected to be paid off, the project company will pay 20 per cent of the profits to the Turkish government. The agreement also provides for setting up a fuel fabrication plant in Turkey. It is reported that Turkey's energy ministry plans to draft a new nuclear energy bill which will be brought to parliament for ratification in June 2010 that will establish the Akkuyu plant as a 'state to state partnership', without holding an open tender. The bill may also cover a second planned nuclear plant at Sinop on the Black Sea. In March 2010 an agreement was signed with Korea Electric Power Corporation (KEPCO) permitting it to prepare a bid for four APR-1400 reactors (WNA 2010b).

The latest arrangements may resolve the perpetual problems Turkey has had in its financing and tendering processes for nuclear power. But they still cannot guarantee that Turkey's notoriously byzantine and slow political decision-making processes will ultimately lead to actual deployment of nuclear power by 2030.

The United Arab Emirates

The UAE announced in 2008 that it would pursue a nuclear energy programme (Salama 2008). It currently hopes to have three 1,500 MWe reactors running by 2020, accounting for 15 per cent of its energy needs (WNA 2010b; Hamid 2008). Although the UAE has the world's sixth largest proven oil reserves and fifth largest proven natural gas reserves (CIA 2009), it has been making a strong economic case for nuclear power based on its rapid economic growth and a predicted fall in natural gas production (*Gulf News* 2008). By generating electricity using nuclear power, the UAE can also export more oil and natural gas instead of using it for domestic consumption (WNN 2008j). The UAE has an existing electrical grid capacity of approximately 16,000 MW, but analysis has shown that by 2020 peak demand will reach nearly 41,000 MW, a 156 per cent increase in just over a decade (Kumar 2008). Fresh water resources are extremely limited, prompting plans to build a 9,000 MW desalination complex in Dubai that could be powered by nuclear energy (Kessler and Windsor 2007: 124). Some 20 GWe of nuclear power is envisaged from about 14 plants, with nearly a quarter of this capacity operating by 2020. Two sites are envisaged between Abu Dhabi and Qatar, and possibly at Al Fujayrah on the Indian Ocean coast.

Although pundits predicted that the UAE would be 'several decades' away from generating nuclear power because it lacks a sufficient technical and legislative framework (Kessler and Windsor 2007: 130), the federation has moved aggressively to court foreign reactor vendors, sign nuclear cooperation agreements with other countries and hire foreigners, lured by extraordinary salaries, to set up its regulatory authority. The UAE is particularly active in paying others to manage its future nuclear power programme. According to Philippe Pallier, director of France's Agence France Nucléaire International, the UAE is creating a new model for management of a national nuclear power programme, one based on contractor services rather than on indigenous management expertise (MacLachlan 2008b: 6).

The UAE has also sought to be a model new entrant in providing reassurances about its peaceful intentions by concluding an Additional Protocol to its safeguards agreement and renouncing any ambition to enrich uranium or reprocess plutonium. It has concluded a so-called 123 Agreement with the United States, which entered into force in late 2009 and which provides additional legal assurances (Ramavarman 2009).

Institutionally, the UAE has established a Nuclear Energy Program Implementation Organization which has set up the Emirates Nuclear Energy Corporation (ENEC) as a public entity, initially funded with $100 million, to evaluate and implement nuclear power plans within the Emirates reactor (WNA 2010b). In October 2009 the Federal Law Regarding the Peaceful Uses of Nuclear Energy was signed, providing for development of a system for licensing and control of nuclear material, as well as establishing the independent Federal Authority of Nuclear Regulation to oversee the UAE's nuclear energy sector, headed by a senior US regulator. The law makes it illegal for the UAE to develop, construct or operate uranium enrichment or spent fuel processing facilities.

In December 2009 the UAE announced that it had accepted a bid from a consortium led by South Korea's KEPCO for four APR-1400 reactors to be operational by 2020. The value of the contract for the construction, commissioning and fuel loads for four units is about $20 billion, with a high percentage of the contract being offered under a fixed-price arrangement. The consortium expects to earn another $20 billion by jointly operating the reactors for 60 years. The firm C2HM Hill has been contracted to provide 'full service' management, engineering, construction and operations for the UAE's new build (WNA 2010b). The site selected is at Braka, between Qatar and Abu Dhabi.

While finance is unlikely to be the obstacle that it is in other developing countries, the financial crisis that hit Dubai in November 2009, as well as fluctuating oil revenues, serve as a reminder that not even oil-rich countries are immune from the travails of nuclear economics. The country may have difficulty meeting its projected energy demand to 2030 using nuclear power, but it is one of the more likely countries to succeed in its long-term development of a nuclear power industry.

Vietnam

In May 2001, the government instructed the Ministry of Industry (MOI), assisted by the Ministry of Science and Technology (MOST), to conduct a 'pre-feasibility study' examining the prospect of establishing a nuclear power sector (Van Hong and Anh Tuan 2004: 5). Its affirmative report led to the creation of a Nuclear Energy Programme Implementing Organization (NEPIO) in 2002 and the Agency for Radiation Protection and Nuclear Safety Control (VARANSAC) in 2003. The government approved its *Long-term Strategy for Peaceful Utilization of Atomic Energy up to 2020* in January 2006, and took the decision in June 2008 to construct four reactors at two nuclear power plants (WNA 2008l).

In November 2009 the National Assembly gave its approval for the two plants, demonstrating the government's determination to move ahead despite concerns about whether Vietnam can handle the high cost and complexity of the project. In April 2010 it was reported that Russia would build Vietnam's first two nuclear power units, both of 1,000 MW, in central Ninh Thuan province. Construction is scheduled to start on the first unit by 2014 and to be completed by 2020 (*Nucleonics Week* 2010).

Vietnam has a low GDP and a relatively small electricity grid that would seem to militate against nuclear power. Yet others view the country's real GDP growth rate of more than 7 per cent in the past two decades, its quickening industrial development and its authoritarian government as likely to enable it to persist with its plans (Gourley and Stulberg 2009: 6). But as Vu Trong Khanh and Patrick Barta note, the estimated cost of the two plants, around 200 trillion

Vietnamese dong ($11.3 billion) is 'a hefty price tag' when Vietnam has just devalued its currency and faces rising debt payments (Khanh and Barta 2009).

Conclusions

This survey of likely trends in civilian nuclear energy to 2030, both worldwide and on a country-by-country basis clearly presents a mixed picture. Unlike other surveys it does not result in cumulative numbers of reactors likely to be built by 2030 or 'guesstimates' of the likely total contribution of nuclear energy to global electricity production. What it illustrates is how fraught such calculations can be and how care must be taken with extrapolations based on broad indices of economic growth, electricity demand and, perhaps most importantly, announced government intentions and plans. As the country-by-country analysis shows, each current nuclear energy state and each aspirant faces its own particular challenges in either increasing its current capacities or moving to nuclear power for the first time. The developing countries in particular face enormous hurdles in terms of governance, physical infrastructure and finance.

While there is no scientific method for weighing the balance of the drivers and constraints detailed in this study, it is clear that an expansion of nuclear energy worldwide to 2030 faces considerable barriers which, while not insurmountable, are likely to outweigh the drivers of nuclear energy. It is true that there are signs of life in the nuclear power industry that have not been seen since the 1980s, driven by concerns about energy security and climate change and by a growing demand for electricity worldwide. Scores of states, including developing countries, have expressed interest in nuclear energy and some have announced plans. Several existing nuclear energy states, notably in Asia, are already building new reactors, while others are studying the possibilities. There is certainly a revival of interest.

Yet globally, while the gross amount of nuclear-generated electricity may rise, the percentage of electricity contributed by nuclear power is likely to fall as other cheaper, more quickly deployed alternatives come online. An increase as high as a doubling of the existing reactor fleet as envisaged in some official scenarios seems especially implausible, given that it can take a decade of planning, regulatory processes, construction and testing before a reactor can produce electricity. While the numbers of nuclear reactors will probably rise from the current number, the addition of new reactors is likely to be offset by the retirement of older plants, notwithstanding upgrades and life extensions to some older facilities.

The economics of nuclear power are the single most important constraint and these appear to be worsening rather than improving, especially as a result of the recent global financial and economic turmoil. Private investors are wary of the high risk, while cash-strapped governments are unlikely to provide sufficient subsidies to make even the first new build economic. Developing countries will, by and large, simply be priced out of the nuclear energy market. The pricing of carbon through taxes and/or a cap-and-trade mechanism will improve the economics of new nuclear build compared with coal and gas, but will also favour less risky alternatives like conservation, energy efficiency, carbon sequestration efforts and renewables. Nuclear is not nimble enough to meet the threat of climate change. Demand for energy efficiency is leading to a fundamental rethinking of how electricity is generated and distributed that will not be favourable to nuclear.

The nuclear waste issue, unresolved 60 years after nuclear electricity was first generated, remains in the public consciousness as a lingering concern. Fears about safety, security and nuclear weapons proliferation also act as dampeners of a nuclear revival. In short, despite some powerful drivers and clear advantages, a revival of nuclear energy faces too many barriers compared to other means of generating electricity for it to capture a growing market share in the coming decades.

In terms of technology, most new build in the coming two decades is likely to be principally in the form of third-generation light water reactors, using systems that are expected to be more efficient, safer and more proliferation-resistant, but not revolutionary. Nuclear power will continue to prove most useful for baseload electricity in countries with extensive, established grids. Large nuclear plants will continue to be infeasible for most developing states with small or fragile electricity systems. Small reactors will not be deployable in sufficient numbers to make much of a difference to overall electricity supply given the current unknowns concerning price, reliability, safety, security and proliferation-resistance. Generation IV systems will not be ready in time, and nuclear fusion is simply out of the question. The 'plutonium economy', even in determined countries like India and Japan, is likely to remain a dream.

It is thus likely that the nuclear energy 'revival' to 2030 will be confined to existing nuclear energy producers in East and South Asia (China, Japan, South Korea and India); Europe (Finland, France, Russia and the UK); and the Americas (Argentina, Brazil and the United States). One or two additional European states may adopt or return to nuclear energy, notably Italy, Poland and Turkey. A smaller number of developing states, those with oil wealth and command economies, may be able to embark on a modest programme of one or two reactors. The most likely candidates in this category appear to be Egypt, Indonesia, Jordan, the UAE and Vietnam, although all face significant challenges in achieving their goals. For the vast majority of aspiring states, nuclear energy will remain as elusive as ever.

4   The current status of global nuclear governance

The nuclear safety regime

The 1986 Chernobyl disaster was a 'wake-up call' to the nuclear industry, national governments and the international community about the potential transboundary effects of nuclear disasters. It demonstrated the truism that global nuclear safety requires a global, not purely national, approach. It led, in record time, to the emergence of a true international, legally-binding nuclear safety regime where hitherto there had only been a patchwork of largely voluntary arrangements. Numerous other initiatives were taken by industry, government and international bodies to strengthen global governance of nuclear safety.

The regime remains, however, sprawling and loosely integrated, reflecting its episodic and largely uncoordinated evolution. It comprises legally binding international conventions; non-binding international safety standards; programmes to facilitate the implementation of those standards by international organizations and multinational networks; the efforts of the international nuclear industry itself; and the activities of the national nuclear infrastructure of each state, including venders, operators and regulators.

Although serious accidents and safety breaches in any part of the civilian nuclear industry have implications for the reputation of the civilian nuclear industry generally, the focus here will be on the safety of nuclear reactors, since these are central to the nuclear revival and are an important concern for the public. Moreover, nuclear reactors are considered most at risk of a serious accident because they are designed to operate in a state of controlled criticality (Nuttall 2005: 37) and because a severe accident may release radioactivity not just locally but via atmospheric transport across a wide area, including over national borders.

Nuclear reactor safety

The risk of a major nuclear power plant accident is generally judged to be low, although the precise risk is highly contested.[1] Since the advent of civilian nuclear energy in the 1950s there have been eight nuclear reactor incidents resulting in damage to the reactor and release of radioactivity, including that at Three Mile Island in Pennsylvania in 1979.[2] Only one has resulted in a major release of radioactivity with significant environmental consequences – the 1986 Chernobyl accident in the Ukraine. In addition, there have been several other known accidents (called 'precursor events' in the United States) in which nuclear reactor systems malfunctioned but there was no release of radioactivity. These include the Unit 1 fire at the Browns Ferry reactor in Alabama in 1975.[3] More recently, in 2002 there was a 'near miss' incident at the Davis-Besse plant in Ohio, in which boric acid leaking from inside the core ate a pineapple-size hole through the carbon steel top of the reactor vessel (Keystone Center 2007: 174–175). The problem that the nuclear power industry faces is that while the probability of a severe nuclear reactor accident is low, such an event would have enormous environmental, economic and political consequences.

Nuclear safety has improved in many states and in many areas since Three Mile Island and Chernobyl. All of the Chernobyl-style Soviet reactors have been shut down, while other Soviet reactors have had safety upgrades. Independent international safety reviews have identified significant progress in Eastern European countries to improve the safety of their nuclear power plants since the early 1990s (Trosman 2009: 65). Nuclear safety has also improved worldwide. According to WANO, the three most important plant-based Performance Indicators of improved safety – the rate of unplanned 'automatic trips' (when the reactor is automatically shut down by safety systems rather than plant operators), radiation exposure of workers and discharges to the environment have 'drastically decreased' compared with 40 years ago when widespread use of civilian nuclear energy began (WANO 2009; NEA 2008: 224). On average, according to the NEA, worker exposures at operating nuclear power plants halved between 1992 and 2006. Such gains were relatively uniform across all types of reactors (NEA 2008: 219).

In March 2009, on the thirtieth anniversary of Three Mile Island, the Nuclear Regulatory Commission (NRC) reported that the number of significant reactor events in the United States – those with serious safety implications such as a degraded fuel rod – had dropped to nearly zero over the past 20 years (NRC 2009a). The average number of times that safety systems have had to be activated is about one-tenth what it was 22 years ago. Radiation exposure of plant workers has steadily decreased to about one-sixth of 1985 levels, well below US federal government limits. The average number of unplanned annual reactor shutdowns decreased nearly ten-fold (to 52 in 2007 from 530 in 1985). Improvements in other aspects of nuclear safety, especially ephemerals such as safety culture (a major factor in the Davis-Besse incident (Lochbaum 2006: 38)) are, however, less easily measured and hence more contestable.

The NEA claims that a major reason for improved safety performance has been the extensive use of lessons learned from operating experience (NEA 2008: 226). This has resulted in the retrofitting of safety systems, improved operator training and emergency procedures and a focus on human factors, safety culture and quality management. The philosophy of continuous safety improvement has, according to the NEA, been adopted by many in the nuclear industry. But there are limits, as embodied in the industry's 'as low as reasonably achievable' (ALARA) principle (NEA 2008: 212). As in any mature industry there is an 'asymptotic' limit to future major improvements given the cost–benefit trade-offs necessary in seeking safety perfection.

But as INSAG Chair Richard Meserve notes, 'noteworthy safety lapses continue to occur at nuclear power plants around the globe, including at reactors in countries with extensive operational experience and strong regulatory capabilities' (Meserve 2009: 102). A study by the Union of Concerned Scientists has revealed that in the 27 years since Three Mile Island, 38 US nuclear power reactors have had to be shut down for at least one year while safety margins were restored to minimally acceptable levels (Lochbaum 2006). Seven of the reactors experienced two-year outages, a majority caused not by broken parts but by a 'general degrading of components to the point that safe operation of the plant required a shutdown for broad, system-wide maintenance' (Lochbaum 2006).

In July 2008 the US Government Accountability Office (GAO), after studying just ten US reactors, reported that some of the country's nuclear power plants have yet to comply with some of the government's fire safety regulations issued after the 1975 Browns Ferry fire (GAO 2008). That event was caused when a worker using a candle to check for air leaks ignited electrical cables. Longstanding unresolved issues include continuing reliance on manual actions by plant workers to ensure fire safety (for example a worker turning a valve to operate a water pump), rather than 'passive' measures (such as fire barriers and automatic fire detection and suppression). As for the NRC itself, it has no centralized database on the status of compliance. The GAO noted that while the recommended fire standards were being adopted for half of the US reactor fleet, the operators doing so faced significant 'human capital, cost and methodological challenges', including a 'lack of people with fire modeling, risk assessment and plant-specific expertise' (WNN 2008b; GAO 2008).

Such difficulties are not confined to the United States. France's Nuclear Safety Authority (ASN) wrote to Areva CEO Anne Lauvergeon in August 2009 to ask the company to undertake a 'broad review of safety management' across all divisions and subsidiaries after several incidents at its fuel cycle facilities (MacLachlan 2008a: 3). The regulator's Groupe Permanent D'Experts had reportedly never met or done a thorough review of safety management across the Areva group.

Nuclear safety is thus, necessarily, a work-in-progress, particularly in terms of the human dimension that is difficult to 'engineer' out of the system. Then NRC Chair Dale Klein told his agency's 21st Annual Regulatory Conference in March 2009 that 'We have continued to see incidents over the last few years … that indicate that safety culture was not a priority through all the staff at all the plants' (Weil 2009a: 1). Paradoxically, even a strong performance record can lead to problems due to complacency.

Nuclear safety is relevant to the entire civilian nuclear fuel cycle, including uranium mining and milling, uranium conversion and enrichment facilities, fabrication plants and reprocessing facilities. It also applies to nuclear transport and nuclear waste storage facilities, both temporary and permanent. The issue of nuclear safety is also pertinent to the entire life-cycle of nuclear facilities, including their design, construction, operation, startup, shutdown, maintenance, decommissioning, dismantlement, site cleanup and disposition of contaminated materials.

The Convention on Nuclear Safety

The 1994 Convention on Nuclear Safety (CNS) is the most important legally-binding instrument in the nuclear safety field. It was adopted in June 1994, opened for signature in September 1994 and entered into force in October 1996. As of September 2009 there were 66 contracting parties and 13 signatories (IAEA 2009c), including all states that currently have operating nuclear power plants or are building them – with the significant exception of Iran.

The treaty applies to land-based civilian nuclear power reactors, including existing, decommissioned and (importantly in terms of a nuclear revival) future plants. It also covers the generation of radioactive waste by such a nuclear installation and any related treatment and storage of spent fuel and waste on the same site (IAEA 1999: Art. 19). This is significant in that most countries store radioactive waste or spent fuel at nuclear power plants pending the opening of long-term geological or other disposal sites.

However, the CNS excludes other nuclear fuel cycle facilities for fuel fabrication, uranium conversion and enrichment, and reprocessing. This represents a worrying lacuna in the nuclear safety regime. As the IAEA points out, 'Fuel cycle facilities face unique nuclear safety challenges such as criticality control, chemical hazards and susceptibility to fires and explosions' (IAEA 2009j: 16). In March 2006, for instance, there was a near criticality accident involving highly enriched uranyl nitrate at a facility involved in down-blending HEU to LEU in Erwin, Tennessee. The plant has reportedly had a history of 'regulatory challenges' and 'ineffective solutions' (Horner 2008: 8). In France there were incidents at two of Areva's fuel cycle facilities in July 2008, the uranium waste treatment plant at Tricastin and the Cerca research reactor fuel fabrication facility in Romans (MacLachlan 2008a: 3–4). Many fuel cycle facilities rely heavily on operator intervention and administrative controls to ensure nuclear safety, rather than the gamut of approaches applied to nuclear power plants.

Acknowledging in its preamble that 'responsibility for nuclear safety rests with the State', the CNS sets out an international safety 'framework' within which states should operate. The preamble also declares that parties only commit themselves to the application of 'fundamental safety principles' rather than 'detailed safety standards'. As the negotiating history shows, these principles derive 'to a large extent', although with some weakening, from the IAEA's 1993 Safety Fundamentals document 'The Safety of Nuclear Installations' (IAEA 1993). Although the CNS does not require compliance with the IAEA's safety standards and guidelines and in fact does not even refer to them, it is clear that these have the greatest global credibility and legitimacy and are thus considered by states to be the international benchmark against which they should measure their compliance with the CNS.

Essentially, though, the CNS is as much about activities and measures as it is about safety principles. It requires each state party to:

- immediately assess the safety of existing reactors and if necessary effect improvements or shut them down;
- take the necessary legislative, regulatory and administrative steps to implement their obligations under the convention, including: national safety regulations; a licensing system for nuclear installations, an inspection and assessment system; and sanctions in the event of breaches;
- establish domestic legal provisions that at a minimum mirror those in the treaty;
- establish a regulatory body with the necessary authority, competence and resources;
- conduct a comprehensive and systematic safety assessment prior to a nuclear plant being allowed to operate and repeat this exercise periodically throughout its lifetime;
- undertake verification activities to ensure the safe operation of all installations using analysis, surveillance, testing or inspection;
- put in place emergency plans, both on-site and off-site, to mitigate the consequences of any radiation release;
- ensure that installation design and construction provide for 'defence in depth' against the release of radioactive materials;
- ensure that relevant levels of maintenance, inspection and testing are conducted by plant operators and that procedures exist to respond to operational incidents and accidents; and
- ensure that safety-related engineering and technical support is available and that all significant safety incidents are reported.

The CNS has no monitoring, verification or compliance system and no penalties for non-compliance. The convention's preamble vaguely describes it as an 'incentive instrument', although it is not clear how this differs from other treaties, most of which contain incentives of some type. Instead of verification, the parties committed themselves to peer review – at the time a significant innovation in nuclear governance. Peer review entails each party providing all others with a detailed periodic report on the measures it takes to implement the convention. Review meetings are convened every three years to review such reports, with states usually represented by their national regulators. The texts are submitted six months in advance and circulated to all contracting parties for written exchanges of questions, answers and comments. Unusually in international agreements, attendance at such meetings is mandatory.

Given the concern of Western European states about the safety of Soviet-type nuclear reactors after Chernobyl, most attention in the early days of the convention fell on the requirement that states immediately assess the safety of existing reactors and if necessary effect improvements or shut them down. Shutdowns did occur in the former East Germany, Bulgaria, Lithuania and Slovakia, in large part due to pressure from the then Group of 7 (G7)[4] and the incentive of accession to the European Union. After a lacklustre report to the first CNS review meeting, Russia was pressured to provide to the second meeting a more convincing account of the measures taken to install safety retrofits to its own Soviet-era reactors. It reportedly did this to the satisfaction of its treaty partners.

The most recent review meeting (the fourth) was held in April 2008 (IAEA 2008m). Fifty-seven of the 61 parties (93 per cent) submitted national reports. Compliance with the reporting requirements was therefore excellent. The four parties that did not submit a national report in 2008 were Kuwait, Mali, Nigeria and Sri Lanka. Six states – Bangladesh, Kuwait, Mali, Moldova, Sri Lanka and Uruguay – did not attend the meeting despite their legal obligation to do so. None of these non-compliant states have nuclear power plants, although at least three, Bangladesh, Kuwait and Nigeria, have declared their interest in acquiring them.

The summary report agreed by the contracting parties concluded that the national reports were in many cases of high quality and provided ample information (IAEA 2008m: 15–16). A 'high degree of compliance' was reported (IAEA 2008m: para. 3). The discussion of national reports apparently 'resulted in identification of good practices, challenges and planned measures to improve safety'. In general, the report claimed, 'the overall safety and radiation protection performance' at nuclear power plants 'appear to remain satisfactory'. The parties themselves conceded, however, that in making judgements about compliance with the CNS they are forced to 'rely on the accuracy and completeness of the information provided by each contracting party and in its answers to the questions asked of it' (IAEA 2008m: para. 22). They also cautioned that 'the worldwide nuclear industry and regulators must avoid complacency' (IAEA 2008m: para. 3).

Drawing conclusions about the reality of compliance with the CNS – or more pertinently the reality of nuclear reactor safety in each country – based solely on the CNS peer review system, is problematic. The reports are not meant to be an assessment of the level of nuclear safety per se, but rather an account of the measures that each country has put in place to help implement the convention. To what extent such measures are effective or whether, in the worst case, they are mere window-dressing is open to question. On the issue of safety culture, increasingly recognized as one of the lynchpins of nuclear safety but difficult to measure, the summary report noted that it is now 'in place' in only some state parties, implying that it is not yet mature or commonplace (IAEA 2008m: para. 25).

Yet the peer review process is clearly effective in exposing the parties to critical scrutiny. The national reports are examined carefully, detailed questions are asked in advance, and during the question-and-answer sessions there is reportedly polite but pointed, and at times persistent, probing. Only security-related issues are off-limits. Representatives are pressured not just to provide assurances that problems will be fixed but are expected at the subsequent meeting to provide information on the steps actually taken. The intense peer review can cause unease for representatives of some countries, especially those in Asia, where losing face is culturally taboo, or countries like Russia with a tradition of pervasive state secrecy. Yet all of these countries have presented comprehensive reports and undergone intense questioning about their compliance with the CNS. As familiarity with each country's situation has improved, the process has become increasingly focused on particular issues of concern. As in all peer review processes those doing the reviewing appear to gain as much as those being reviewed.

Procedures have been put in place to avoid the 'inherent danger of under-enforcement' that peer review processes pose (Handl 2003: 19). The sub-group structure, with its mix of states with nuclear reactors and those without them, is designed to help avoid mutual reluctance to criticize peers or political pressures unrelated to nuclear safety. Another safeguard is the random electronic reshuffling of country group membership for each meeting to attenuate potential 'group think'. A growing challenge is information overload. In practice only the major nuclear energy powers will have the time or personnel to analyse each report in detail.

The CNS suffers from a lack of openness and transparency, making it impossible for outsiders to truly assess the system's effectiveness. Interested non-governmental organizations (NGOs), academics or other members of the public are not permitted to attend the proceedings. Parties are encouraged to make their own reports public, but as of October 2009 only 23 were posted on the IAEA website, six less than are available from the 2005 meeting.[5] Proposals have been made for opening up at least some parts of the review meetings to the public but these have not achieved general agreement to date. There is a natural tension and trade-off between confidentiality and transparency, but governments, at least in many Western democracies, have policies and mechanisms that seek to achieve the right balance.[6] Such practices could be emulated by the CNS review meetings. The difficulty will be achieving agreement from states that domestically are unused to such openness.

The parties continue to seek to improve the CNS process by convening open-ended working groups to consider ideas. The 2008 meeting agreed on steps to improve inter-sessional communication and to make the review process more efficient, following the failure of efforts to do so after the 2005 meeting (MacLachlan 2008d: 10). One potentially valuable reform is to hold joint meetings between the parties to the CNS and the parties to the Joint Conventions on the Safety of Spent Fuel Management and on the Safety of Radioactive Waste Management to discuss issues of concern.

Role of the International Atomic Energy Agency

Nuclear safety is one of the three pillars of IAEA activities – in addition to the promotion of nuclear energy and the non-proliferation of nuclear weapons. The Agency's role as the global 'hub' of nuclear safety has been steadily enhanced since the Chernobyl disaster. In addition to becoming the secretariat for all of the new safety-related conventions, its key activities in nuclear safety are the setting and promotion of safety standards, safety advisory missions and management of peer review processes.

Setting and promoting safety standards

The IAEA has created comprehensive, detailed sets of safety standards covering all aspects of the peaceful uses of nuclear energy. The development of such standards is overseen by a Commission on Safety Standards (CSS) and its various safety committees, comprised of senior government officials with national responsibilities for nuclear safety. In order to ensure the broadest possible consensus, safety standards are submitted for approval to the IAEA Board of Governors (ElBaradei 2003a: v).[7] In addition to its own experts, the Agency relies on member states, industry and academia. Formal advisory bodies include the International Nuclear Safety Group (INSAG), those convened by the OECD's NEA or the European Atomic Energy Community (Euratom), the European Nuclear Energy Forum and the G8 Nuclear Safety and Security Group (NSSG). INSAG is a group of high-level experts, appointed by the IAEA Director-General, which provides advice not just to the IAEA but to the international nuclear community generally (INSAG 2006b: 3). Its reports are published as IAEA documents. For radiological standards[8] the IAEA draws on the work of the International Commission on Radiological Protection (ICRP) and the United Nations Scientific Committee on the Effects of Atomic Radiation (UNSCEAR).

In respect of civilian nuclear power, IAEA safety standards cover the establishment of legislative and regulatory infrastructure, radiation protection, reactor site evaluation, and the design, safe operation and safe decommissioning of nuclear power plants, as well as nuclear transport. The three levels of IAEA safety documents are:

- Safety Fundamentals, which set out basic objectives, concepts and principles;
- Safety Requirements, which establish basic requirements that 'shall' be fulfilled in the case of particular activities or applications; and
- Safety Guides, which contain recommendations, based on international experience that 'should' be followed in fulfilling the Safety Requirements.

The Agency also establishes guidelines and codes of conduct, such as its 1998 Guidelines for the Management of Plutonium and its 2004 Code of Conduct on the Safety of Research Reactors.

IAEA safety standards are legally binding on the IAEA itself in its own operations and on states in relation to operations assisted by the IAEA. The CNS, along with other international nuclear safety conventions, currently set out general legally-binding undertakings and general safety principles but do not legally oblige states to implement IAEA standards. The question arises whether such standards should be made legally binding and compliance with them verified by international inspectors as in the case of nuclear safeguards. In his last speech to the United Nations General Assembly before his retirement in December 2009 IAEA Director-General Mohamed ElBaradei went so far as to call for IAEA safety standards to be 'accepted by all countries and, ideally, made binding' (ElBaradei 2009), although whether binding in international law or national legislation he did not make clear.

While superficially appealing and logical, it is not clear that making standards legally binding would help, even if it were politically possible. They are arrived at through a consultative process among states and are increasingly recognized as essential, so there is peer pressure to comply. They are also subject to periodic revision based on experience, they are not all applicable to all types of existing, let alone future reactors, and they often are open to legitimate interpretation in their application. Each country's safety regime must fit its own national legal, economic and cultural circumstances if it is to be truly effective and must primarily be the responsibility of the national regulator. Moreover, it is unlikely that compliance would be any greater without the addition of an enforcement mechanism, which states would likely oppose. Finally, an egregious safety record is less likely to be due to wilful intent than a lack of government attention to the problem, poor national governance generally, substandard technical or institutional capacity and insufficient funds. All of these challenges are better solved with international technical assistance than international enforcement.

Moreover, while otherwise not legally binding on IAEA member states or on parties to any treaty, the degree to which national safety requirements are expected to be in compliance with the IAEA Safety Standards depends 'on the level of the publication in the hierarchy' (INSAG 2006a: 11). Safety Fundamentals 'should not be amenable to significant changes over time, and they are expected to be met without exception'. Safety Requirements 'should be met by new facilities and related new facilities, and are a target that should be met over a period of time that is reasonable for existing facilities and practices'. Safety Guides 'are practical guidance on achieving state-of-the-art nuclear safety'; meeting them is 'recommended unless alternative means can be taken to provide the same level of safety'.

Most relevant to a nuclear energy revival is the IAEA's Nuclear Safety Standards (NUSS) programme. When the NUSS documents were re-examined in 1979 following Three Mile Island, it was concluded that the accident did not invalidate any of them and that the IAEA 'had shown foresight in setting up the NUSS program, providing a good basis for the safety of nuclear power plants' (Gonzáles 2002: 285). According to Argentinean nuclear safety expert Abel Gonzáles, NUSS documents 'serve as advisory documents for designers, operators, and regulators, allowing them to check their relevant activities against what is internationally considered to be good practice', but are not expected to tell designers how to design plants or operators how to operate their plants (Gonzáles 2002: 286).

The reason why states have been resistant to making IAEA safety standards legally binding is partly due to their differing reactor technologies and regulatory systems, but also partly due to two competing philosophies about nuclear regulation. One school of thought favours the 'prescriptive approach', setting standards and making compliance with them compulsory, as long favoured by the US Nuclear Regulatory Commission. The second approach is a performance-based one, favoured by Canada and the UK, which sets basic standards and expectations but is flexible about how these are achieved as long as safety is maintained. The NRC has recently indicated that it would move to a more performance-based approach, with greater emphasis on higher-level safety principles and fundamentals. Despite differences in philosophical approaches to nuclear safety most countries appear to support the IAEA role in setting international standards and in providing guidance, advice and assistance in implementing them.

Assistance, services and other activities

The IAEA provides significant advice and assistance to member states on nuclear and radiological safety. First, it publishes a staggering array of publications and facilitates other types of information-sharing.[9] This includes supporting regional networked databases to facilitate regional knowledge-sharing and capacity-building, notably the Asian Nuclear Safety Network (ANSN) and the Ibero-American Nuclear and Radiation Safety Network.

Second, the Agency provides technical assistance, training and services through its Technical Cooperation (TC) programme. In 2009 there were 178 projects on nuclear safety worth almost €26 million (IAEA 2009u). The provision of nuclear safety and radiological services to states has become a major part of the IAEA's nuclear safety agenda. The most important of these for nuclear reactor safety are considered in detail below but they also include: Safety Culture Assessment Review Teams (SCART); International Regulatory Review Teams (IRRT); Engineering Safety Review Services (ENSARS); International Probabilistic Safety Assessment Review Teams (IPSAER); Review of Accident Management Programmes (RAMP); the Transport Safety Appraisal Service (TransSAS); and various radioactive waste management services. Of particular relevance to the nuclear energy revival is the 'Integrated Strategy for Assisting Member States in Establishing/Strengthening Their Nuclear Safety Infrastructure' (IAEA 1997: 3–4), involving a joint review by the IAEA and the state, including identification of areas where safety falls short of the reference situation and where assistance could be most effectively applied.

The Operational Safety Review Teams (OSART) programme is designed to aid states in improving the operational safety of their nuclear power plants, essentially through a process of peer review. Upon request, teams of international experts conduct three-week intensive reviews of a nuclear facility, covering: management goals and practices, organization and administration, training and qualifications of personnel, operations, maintenance, technical support, operational experience feedback, radiation protection, chemistry and emergency planning and preparedness. The programme allows nuclear experts and power plant operators from one country to assist power plant operators in another through the sharing of information and international best practice. An important aspect is to identify strengths that can be shared with other states and fed back into the Agency's work and a comprehensive Agency database, to improve safety standards. After the initial visit a follow-up review one year to 18 months after the initial mission is conducted. The first OSART mission was to the Ko-Ri nuclear power plant in South Korea in August 1983. Since then there have been more than 132 missions, carried out at 87 nuclear power plants in 31 countries (IAEA 2005: 2).

The outcome of missions is typically good, with most operators scoring high grades for safety performance. A mission usually yields between 20 and 30 recommendations and between 40 and 50 per cent of issues are resolved by the operator within a year, with satisfactory progress eventually being made for 96 to 97 per cent.[10] Many host countries and host plants post the OSART reports on their websites to enhance transparency (IAEA 2005: 9). Only three countries with operational power reactors – Armenia, India and Taiwan – have not hosted an OSART mission so far. The OSART programme complements the national peer review process that CNS parties undergo at their review meetings.

Peer Review of the effectiveness of the Operational Safety Performance Experience Review (PROSPER) provides advice and assistance in developing and managing the operational experience feedback process. A mission visits a reactor operator to assess management practices, policies and procedures, the comprehensiveness of instructions, the adequacy of resources, and the overall capability and reliability of personnel (IAEA 2003c). If the feedback process does not meet with internationally accepted best practice, improvements are suggested, with findings and corrective actions reported to the national regulatory body. A follow-up mission, at the request of the state, is conducted within 18 months to assess progress. No details are publicly available on which states have participated in PROSPER.

The IAEA's Integrated Regulatory Review Service (IRRS) provides advice and assistance to enhance the effectiveness of regulatory infrastructure for both safety and security. Importantly, it requires the state to first provide a self-assessment of how, in regulatory terms, it is complying with the CNS and the 1997 Joint Convention on the Safety of Spent Fuel Management and on the Safety of Radioactive Waste Management. These reports are subject to extensive peer review, providing the opportunity for 'open and frank discussions on trends, challenges and best practices' (IAEA 2009j). The requesting state, however, decides on the scope, which may range from a discrete regulatory issue to an entire regulatory enterprise. The process includes site visits, interviews and documentation review. Between 2006 and 2009 the legislative assistance programme reviewed the national laws of 51 countries, more than half of which were African (IAEA 2009h: 20).

Periodic Safety Reviews (PSR), which aim to ensure a high level of safety throughout a nuclear plant's operating lifetime, are conducted by operators and reviewed by the national regulator. The Agency, which recommends that PSRs be conducted every ten years, may be invited to review the conduct of a PSR, which are seen as additional to routine reviews of plant operation and special reviews following major events of safety significance (IAEA 2003a: 1). They aim to assess the cumulative effects of plant ageing and modifications, operating experience, technical developments and siting aspects. The reviews include an assessment of plant design and operation against current safety standards and practices.

The IAEA/NEA Incident Reporting System (IRS) collects information from participating states' national regulators on unusual events in nuclear power plants that may have safety or accident prevention implications. The information is assessed, analysed and fed back to operators to prevent similar occurrences at other plants. The IRS is also concerned with identifying 'precursors', events of apparently low safety significance, which, if not properly attended to, have the potential to escalate into more serious incidents.

Currently all 31 countries that operate nuclear reactors, plus Italy, are participants. While some are active in reporting to the IRS, others never report. In 2006 the IRS received just 80 reports, compared to 1,000 for a reporting system operated by the World Association of Nuclear Operators. The Chairman of INSAG, Richard Meserve, has suggested that regulators are not reporting enough incidents or providing enough information on how they have used others' operating experience (MacLachlan 2007a: 10). In fact the failure of states to report and share experience could be regarded as non-compliance with the CNS, which requires parties to 'take the appropriate steps to ensure that ... existing mechanisms are used to share important experience with international bodies and with other operating organizations and regulatory bodies' (IAEA 1994: article 19(vii)).

Fuel cycle services

Although the CNS does not cover fuel cycle facilities the IAEA does offer member states some safety services for them. There has recently been increasing openness among operators of fuel cycle facilities to share safety information and more use is being made of the Fuel Incident Notification and Analysis System (FINAS) developed by the IAEA and the OECD/NEA (IAEA 2009j: 16). The Agency also offers a Safety Evaluation During Operation of Fuel Cycle Facilities (SEDO) service to assist member states, at their request, in enhancing safety at their fuel cycle facilities. It is a peer review process that bases its performance evaluation on IAEA safety standards and the expertise of its team. Its objective is to promote the continuous development of operational safety and the dissemination of information on good safety practices at fuel cycle facilities. However it does not systematically evaluate and enhance nuclear safety measures. The Agency says it is continuing its efforts to establish a complete set of safety standards to cover all types of fuel cycle facilities (IAEA 2009j: 16). As in the case of nuclear weapons-related facilities, accidents at commercial fuel cycle facilities can taint the prospects for the revival of nuclear energy. Former NRC

Chairman Dale Klein suggested that national nuclear programmes would benefit from 'more formal mechanisms' for cooperating in 'overseeing the nuclear fuel cycle' (Nuclear News Flashes 2007b; Klein 2007).

The International Seismic Safety Centre (ISSC) is a focal point on seismic safety for nuclear installations (IAEA 2009j: 14), assisting states in assessing seismic hazards faced by nuclear facilities in order to mitigate the consequences of earthquakes. The IAEA has begun revaluating the integrity of existing nuclear installations, taking into account recent earthquakes, such as those in Japan which led to plant shutdowns, and other extreme natural events such as tsunamis (IAEA 2009j: 9). This is highly relevant to the 'nuclear revival' as a number of putative nuclear energy states, such as Indonesia and Turkey, are in active seismic zones.

Other international bodies involved in nuclear safety

Several other bodies besides the IAEA, both governmental and non-governmental, are involved in nuclear safety, with some degree of cooperation and collaboration between them. Most notable is the collaboration between the OECD/NEA and the IAEA. The NEA works closely with the Agency on several projects, such as the Multinational Design Evaluation Program (MDEP) for harmonizing regulatory approaches to new reactor designs. The non-governmental bodies, on the other hand, tend to keep a diplomatic distance from the IAEA and each other.

The World Association of Nuclear Operators (WANO)

WANO was established at a meeting in Moscow in 1989, in direct response to the Chernobyl disaster, to enhance the safety of nuclear plants worldwide. It does so by facilitating 'communication, comparison and emulation' among its members in order to maximize safety and reliability. Headquartered in London, it has four semi-autonomous branches in Atlanta, Paris, Moscow and Tokyo. It sees itself as complementary to national and international regulators and does not advocate nuclear power or particular nuclear policies (Crawford 2009). Members sign a confidentiality statement, which limits the ability of outsiders to assess WANO's effectiveness and that of its 'member' nuclear reactors.

Membership of WANO is open to all companies that operate electricity-producing nuclear power plants and organizations representing nuclear operators. While WANO claims that 'Every single organization in the world that operates a nuclear electricity generating power plant has chosen to be a member of WANO', in fact its membership of more than 30 is a mix of individual operating companies and national organizations that represent operators. Thus the United States is represented by the Institute of Nuclear Power Operators (INPO). Nonetheless, impressively, all operators of nuclear power plants are directly or indirectly represented. This includes the Atomic Energy Organisation of Iran (which does not yet have a functioning power reactor), the Nuclear Power

Corporation of India Ltd, and the Pakistan Atomic Energy Commission. In 2006 the British Nuclear Group Sellafield became the first operator of a reprocessing facility to join WANO.

WANO, like the IAEA, runs a peer review system. In 2008 it conducted reviews at 29 nuclear power plants, bringing its total to 387 since the programme began in 1992 (IAEA 2009j: 40). As of 2009 all operating reactors had had at least one peer review and 70 per cent of WANO member 'stations' (120) had hosted two or more peer reviews since the programme began.[11] The 2007 WANO Review (its planning document) established a long-term goal of at least one peer review every six years at each reactor, although acknowledging that more frequent reviews may be necessary. There is currently no systematic peer review of the safety of the rest of the civilian nuclear fuel cycle, but Sellafield's membership of WANO and acceptance of a peer review may help lead to this in future.

WANO Chairman William Cavanaugh III warned WANO members in 2007 that many of them are not assimilating the recommendations from peer reviews. Analysis by WANO staff showed that 'the most common and significant weaknesses in plant performance are similar to those already identified in previous years ... The success of the peer review program is being tainted by issues only being resolved at the symptom level' (Weil 2007: 14). Like those of the IAEA's OSART programme, the results of WANO's peer reviews are confidential (not even shared with the IAEA) and do not carry the same weight as those conducted under an authoritative international body like the IAEA.

In addition to peer reviews and professional development through workshops, seminars, expert meetings and training courses, WANO conducts over 200 technical support missions each year, where a group of highly qualified peers visits a plant to solve a specific issue. WANO also oversees two types of operational reporting: Operating Experience Reports (OER) and Significant Operating Experience Reports (SOER). While the number of events reported to WANO has risen sharply from 321 in 2004 to 936 in 2006, the aggregate numbers belie the great disparity among members, with some members reporting many events and some next to none (Weil 2007: 1, 14). Moreover, the jump in reported events is somewhat artificial, since WANO now includes events with low safety significance.

As noted above, the IAEA runs an incident reporting system, the IRS, which receives reports from regulators rather than plant operators, has different reporting criteria and records far fewer incidents than WANO's. The two systems operate independently and their data is treated as confidential and not shared with non-member organizations or entities, or in the case of WANO, with national regulatory authorities (INSAG 2006a: 14). The latter are excluded due to the risk that they will be forced to reveal the information under national freedom-of-information or other transparency measures (INSAG 2006a: 14). WANO only notifies the IAEA of 'trends'. INSAG Chairman Richard Meserve says there is a 'serious disconnect' between the two systems and a 'need to make data available to international regulators' (MacLachlan 2007a: 10).

As part of its effort to promote the exchange of operating experience, WANO usefully compiles Performance Indicators for safety system performance that are available publicly (WANO 2009). Meanwhile, many of the owner groups for different nuclear plant types have developed experience-sharing networks, but their insights are often limited to the technically unique issues they encounter and they operate under proprietary confidentiality rules (INSAG 2006a: 15).

WANO officials confirm that operators are not reporting all incidents and, as in the case of the IAEA system, are not using others' operating experience to avoid making the same mistakes through its 'lessons-learned' process (Weil 2007: 1, 14). For example, despite many reports of circulating water intake blockage, the frequency of such events worldwide has not lessened. Moreover, the frequency of events concerned with rigging, lifting and material handling has worsened over the six years to 2007. WANO had warned operators not to use foreign materials in order to prevent failures in fuel turbines and generators, but such events continue to occur. Control room culture issues and valve misalignments continue to 'proliferate' and many operators ignore or learn to live with longstanding equipment problems (Weil 2007: 14). Despite all of the measures put in place by the IAEA, WANO and others, INSAG concludes that:

> The OEF [international operating experience feedback] systems available today are not adequate to meet the needs of the ever-increasing number of nuclear stakeholders. There is an acute need to improve the mechanisms that are in place for sharing operating experience, as well as to develop newer, simpler processes to expand on these overtaxed mechanisms. Both the positive (good practices) and the negative (root causes) aspects of OEF must be shared if they are to be effective at reducing and eliminating risks.
>
> (INSAG 2006a: 15)

The Nuclear Energy Agency (NEA)

The NEA's mission is to 'assist its Member countries in maintaining and further developing, through international co-operation, the scientific, technological and legal bases required for the safe, environmentally friendly and economical use of nuclear energy for peaceful purposes' (NEA 2009b). To achieve this it focuses on selected areas and produces authoritative assessments that reflect, or seek to develop, common understandings among member states, notably in the nuclear sciences, safety, regulation, waste management, technical and economic studies, nuclear law and radiation protection. In the safety area, in contrast to the IAEA, the NEA focuses on research and on providing and exchanging information.

The European Atomic Energy Community (Euratom) and the European Commission (EC)

These bodies are not responsible for safety in the European Union countries but help promote it through the cultivation of common views and by identifying best practice. The European Nuclear Safety Regulators Group (ENSREG), established in 2007, is the focal point of cooperation between European regulators and is intended to lead to continuous improvement in nuclear safety, especially in new reactors (IAEA 2009i: 4).

In July 2007 the EC established a European High-Level Group on Nuclear Safety and Waste Management to pursue 'common understandings' and 'reinforce common approaches' in nuclear safety and waste management (Froggatt 2009: 26–27). While intended to lead to binding European nuclear safety standards, including verification of compliance, it failed due to the disparate views among member states about the future of nuclear energy and the need for a common European approach (Ferguson and Reed 2009: 58). The EC directive eventually adopted in June 2009, although legally-binding, only establishes a 'framework' to 'maintain and promote the continuous improvement of nuclear safety and its regulation', but has no provisions for EC verification or EC-wide regulators. Essentially it only requires compliance with the CNS, to which all EU states are already party. States are required to report on implementation of the directive for the first time by July 2014 (which seems rather distant given the importance of the issue) and every three years thereafter, in order to take advantage of the CNS review and reporting cycles. The only novel element is the requirement that at least every ten years member states undertake self-assessments of their national framework and competent regulatory authorities and invite an 'international peer review' of 'relevant segments' which must be reported to member states and the EC.

Apart from WANO, industry-based bodies interested in nuclear safety include the World Nuclear Association (WNA), the World Nuclear Transport Institute (WNTI) and the Institute of Nuclear Materials Management (INMM). Nuclear safety is also one of the concerns of the World Nuclear University (WNU) and the US-based Institute for Nuclear Power Operations (INPO). INPO, established in 1979, nine months after Three Mile Island, has reportedly helped the US industry 'strive for excellence' in plant operations rather than just meet minimum regulatory requirements (Nuclear News Flashes 2009c). Over the past 30 years INPO has reportedly used 'peer pressure, confidential safety assessments, safety inspections, and a principled-based and results-oriented management approach to achieve a high standard of safety while maintaining reliable operations' (Ferguson and Reed 2009: 54). Funded by the US nuclear industry, it sets performance standards and conducts WANO-like plant evaluations that it shares among its members. In 2008 South Korea opened an International Nuclear Safety School (IAEA 2009i: 6), the first in the world.

Role of national nuclear regulators

The role of national nuclear regulators is essential, since they are the channel through which global governance norms, treaty obligations and recommended standards and guidance are implemented nationally. The CNS requires parties to 'ensure an effective separation between the functions of the regulatory body and those of any other body or organization concerned with the promotion or utilization of nuclear energy'. The IAEA's Fundamental Safety Principles, moreover, require that an 'independent regulatory body' be established and sustained. A prominent issue at the 2008 CNS review meeting, prompted by the Canadian government's sacking in January 2008 of the President of the Canadian Nuclear Safety Commission, Linda Keen, was the independence of national regulators. Delegates called the event troubling because of Canada's status as a pioneer in the nuclear power business. Several other states also came under the spotlight because their regulatory bodies were considered too close to organizations that promote nuclear energy, notably those of Brazil, India and South Africa. The meeting agreed to 'further discussion' of the issues.

Global governance arrangements to establish cooperation among nuclear regulators are relatively rudimentary and novel, but growing. Senior regulators meet annually at the IAEA's annual General Conference. While well attended, these last only a day and involve a general discussion on just two themes. However, several regional and reactor-type networks of regulators have arisen to supplement the international regime. They include: the Network of Regulators of Countries with Small Nuclear Programs (NERS); the CANDU[12] Senior Regulators; the Cooperation Forum of State Nuclear Safety Authorities of Countries which operate WWER[13] Reactors; the Western European Regulators Association (WENRA); the Ibero-American Forum of Nuclear Regulators; the NEA Committee on Nuclear Regulatory Activities (CNRA); and the European Nuclear Safety Regulators Group (ENSREG) (IAEA 2009i: 4).

The only international body devoted to regulators that sounds like it is intended to be universal is the International Nuclear Regulators Association (INRA). Established in 1997, it is a small self-nominated 'club' of like-minded senior regulators from Canada, France, Japan, Spain, South Korea, Sweden, the UK and the United States. It operates independently of other international bodies and provides members' regulators with a periodic forum to discuss nuclear safety, notably their collective strategy for multilateral meetings (CNSC 2008: 38). But INRA only includes eight of the 31 national nuclear power plant regulators. A universal international nuclear regulators organization is clearly needed.

Safety of nuclear spent fuel and radioactive waste

Separate from the Convention on Nuclear Safety, but critically important to the safety of civilian nuclear energy, is the part of the global governance regime relating to the safety of spent fuel and radioactive waste. Its goal is to ensure that states and operators of facilities handle spent fuel and radioactive waste safely whether it is in process, being transported, stored or disposed of.[14] The main international agreement is the 1997 Joint Convention on the Safety of Spent Fuel Management and on the Safety of Radioactive Waste Management, while the most important international agency involved is, again, the IAEA.

Joint Convention on the Safety of Spent Fuel Management and on the Safety of Radioactive Waste Management

The objective of the Joint Convention is to achieve and maintain a high level of safety worldwide in spent fuel and radioactive waste management through the enhancement of national measures and international cooperation. Parties must mount effective defences against potential hazards so that individuals, society and the environment are protected now and in the future, prevent accidents with radiological consequences and mitigate their consequences should they occur.

The Joint Convention was adopted and opened for signature in September 1997 and entered into force in June 2001. As of May 2010 there were only 56 state parties plus Euratom and 42 signatories, far fewer than for the CNS (IAEA 2010e). All states with civilian nuclear reactors – with the significant exceptions of India, Mexico and Pakistan, which have not even signed – have ratified the Joint Convention. A couple with significant programmes – China and South Africa – have only recently acceded.

The Joint Convention is the first legal instrument to directly address the major challenges arising from spent fuel and radioactive waste on a global level. The first challenge is that radioactive waste will need to be managed safely well beyond the present generation on a time scale that is 'evolutionary'. Nuclear power generation produces waste that can last more than 10,000 years. The safety of disposal sites for high-level waste across this time span must be independent of institutional control, since no human institutions have ever been known to have lasted that long. Each contracting party must ensure that records regarding the location, design and inventory of the closed facility are preserved and that either active or passive institutional controls remain in place if required. The safety of the disposal site should not rely on such measures (IAEA 2003b: 100).

A second difficulty is that the type of material that some states regard as radioactive waste to be disposed of may be seen by others as an energy resource to be reprocessed for recycling. Hence the Joint Convention deals with both. Many states, implicitly or explicitly, consider that radioactive waste should be disposed of in the state in which it was generated (IAEA 2003b: 97). Most states also consider that whoever was responsible for the generation of waste within the state should be responsible for its safe disposal. Under the Joint Convention ultimate responsibility for ensuring the safety of spent fuel and radioactive waste management rests with the holder of the relevant licence issued by the state regulatory authority (IAEA 2003b: 97). Where there is no such holder, responsibility devolves to the state. State parties are required to incorporate the obligations set out in the convention into their domestic law, having at a minimum, domestic legal provisions that mirror those found in the Joint Convention. A regulatory body must be created that has the authority, competence, and financial and human resources to oversee the safety of waste management and spent fuel management facilities.

The treaty follows the CNS model closely in terms of periodic review meetings, national reports and peer review and these have evolved in similar fashion to those for the CNS. The initial review meetings led to the conclusion that there needs to be a 'holistic' approach to nuclear waste management which 'encompasses all types of radioactive waste from their generation to their reuse, recycling, clearance or disposal' (MacLachlan 2007b: 10).

The Third Review Meeting was held from May 11 to 20, 2009. Forty-five parties participated, including five new parties, China, Nigeria, Senegal, South Africa and Tajikistan. Three parties – Kyrgyzstan, Uruguay and Uzbekistan – failed to attend despite their obligation to do so (although Uruguay did submit a national report). Senegal, Kyrgyzstan and Uzbekistan did not submit a national report as required. None of these states have significant nuclear industries that produce spent fuel or radioactive waste, although Senegal has expressed interest in a nuclear power programme.

In summarizing the results of the Third Review Meeting the contracting parties recognized that there remained 'considerable areas for improvement' (IAEA 2009s: 9). But they also noted that the review process was maturing and that 'more constructive exchanges and more knowledge sharing took place than at previous Review Meetings' (IAEA 2009s: 9). Parties were asked in their reports for the Fourth Review Meeting in 2012 to include or expand their coverage of the following issues (IAEA 2009s: 7): development of a comprehensive legal framework; the effective independence of the regulatory body; implementation of strategies with visible milestones; funding to secure waste management; education and recruitment of competent staff and employees; and geological repositories for high-level waste.

One problem not faced by the CNS national reports is the difficulty of determining exactly what kind of radioactive waste is being referred to in national reports on the Joint Convention. Phil Metcalf, head of the IAEA's Disposable Waste Unit, has noted that diverse terminologies in different countries and even among different facilities in the same country, make communication difficult in the context of the Joint Convention (MacLachlan 2007b: 11). For example, what the Russians call intermediate-level waste, which in Russia has been disposed of in deep boreholes, might qualify in another country as high-level waste that must be packaged and emplaced in a deep geologic repository. After working on new classification guidelines since 1994, the IAEA finally published them in 2008 (IAEA 2008d).

The review meeting process for the Joint Convention appears to operate effectively, in the same manner as that for the CNS, in exposing state parties to probing questions by their peers about their policies and plans in fulfilment of their international obligations. Many delegations are comprised of the same individuals who attend the CNS meetings, effectively making the review process part of a holistic international regime for all aspects of nuclear and radiological safety – as had been advocated by those states, notably the Scandinavian countries, which had wanted the CNS to be comprehensive in the first place. In a sense these countries ultimately achieved their objective, although at the expense of some duplication of organization and process. Given the similarity between the Joint Convention and CNS processes there could be an argument for combining them. This would encourage states themselves to adopt a more holistic approach to nuclear and radiological safety and help avoid past experience where states established nuclear power programmes without giving much, if any, thought to long-term management of the spent fuel and nuclear waste that they were producing or to the long-term costs associated with it.

IAEA safety standards, advisory services and missions

The IAEA became involved in establishing safety standards for radioactive waste soon after its creation in 1957 (Gonzáles 2002: 288). The standards have since then been a work-in-progress as public attitudes evolved and experience was gained in handling such materials. Initially it was envisaged that the radioactive wastes would be disposed at sea and the Agency dutifully drew up regulations to manage this. However, by the 1970s international opinion had shifted to favouring long-term disposition on land in underground repositories and in 1977 the IAEA initiated a programme to produce guidelines on the subject. By the 1980s the growing political salience of the nuclear waste issue induced the IAEA to produce, in 1988, a 'high-profile family of safety standards' (Gonzáles 2002: 289), the Radioactive Waste Safety Standards (RADWASS). In 1995, after being subject to peer review, RADWASS was broadened to include a new emphasis on discharges and environmental restoration, and to rationalize the complex set of Agency documents on the subject.

Today the Agency has a detailed set of safety standards that address radioactive waste and a draft set addressing spent fuel management (IAEA 2009o, 2009t). The regime continues to evolve, especially in the areas of geological disposal and environmental restoration, 'where little or no experience has yet been gained' (Gonzáles 2002: 290). A Working Group on Principles and Criteria for Radioactive Waste Disposal is currently drafting a technical document, *Common Framework for the Disposal of Radioactive Waste* and a safety report, *Model Regulations on Safety of Radioactive Waste Management* (IAEA 2009d).

The Agency's Disposable Waste Unit, which develops the standards that deal with radioactive waste, also assists states in their application. One means is by organizing, again, a peer review, by a team of international experts, who visit to assess and make recommendations regarding the applicable safety standards of the requesting state. Subsequently the IAEA may offer technical assistance to facilitate implementation.

Guidelines for the management of plutonium

One of the little known and remarkable agreements in the area of spent fuel and waste management is the innocuous-sounding Guidelines for the Management of Plutonium (IAEA 2004). In 1992 the IAEA initiated a series of meetings involving countries with the largest plutonium holdings in order to determine the necessity of international methods of managing the fissile material. The countries involved were the nuclear weapon states recognized by the NPT (China, France,

Russia, the UK and the United States), as well as Belgium, Germany, Japan and Switzerland. These countries were concerned about the increasing amounts of civil separated plutonium, as well as the large quantities of fissile material expected to result from the dismantling of nuclear weapons. The proposal thus has its origins in concerns about non-proliferation, safety and security.

In 1993 the IAEA convened an unofficial study of ways to manage plutonium. Participants decided, however, that they preferred to agree in confidence among themselves on such methods, partly to avoid the complications of a long, complicated official negotiation process.[15] The guidelines were agreed in late 1997, communicated to the IAEA in the form of identical letters from the participants and published as an IAEA document in March 1998, along with subsequent declarations of plutonium holdings.

The guidelines express agreement that civil plutonium should be handled in accordance with the major international agreements on non-proliferation, safety, physical protection, material accountancy and control and safeguards, and the rules on international transfers of civil plutonium. The participants also agreed to formulate national strategies on plutonium management that consider the risks of proliferation, especially during storage before irradiation or permanent disposal; the need to protect the environment, workers and the public; and the resource value of the material. Such strategies should also take into account the importance of 'balancing supply and demand' in order to minimize the amount of separated or unirradiated plutonium as soon as is practical.

The most extraordinary aspect of the guidelines, however, is agreement on increasing transparency by publishing the following documents: occasional brief statements explaining the national strategy for nuclear power and spent fuel and general plans for managing plutonium holdings; an annual statement of holdings of all plutonium, subject to the guidelines; and an annual statement of the estimate of the plutonium contained in holdings of spent civil reactor fuel. Such reporting is voluntary since the guidelines are not a legally-binding agreement. However, compliance overall has been good. The submissions are available on the IAEA website (see document INFCIRC/549), and are an unprecedented level of public transparency in this field. According to the Institute for Science and International Security (ISIS) in Washington, DC, however, 'Although the annual publication of civil holdings has been successful overall in creating more transparency, the declarations by several countries are incomplete' (Albright and Barbour 2000). They are also often posted late. Regrettably several states with civilian separated plutonium have not yet chosen to participate, including India, Italy, the Netherlands, Spain and Sweden.

Safety of nuclear transport

International nuclear transport, via air, sea or land, requires by its very nature an international governance regime in a way that no other aspect of nuclear energy generation does.[16] As early as 1959 the United Nations Economic and Social Council charged the IAEA with establishing recommendations on the transport of radioactive material. These were promulgated in 1961 as IAEA Safety Series 6, designed to cover category 7 (covering nuclear and radiological materials) of the hazardous substance identification and classification system set up by the United Nations Committee of Experts on the Transport of Dangerous Goods (IAEA 2003b: 90). This has led to continuing cooperation between the Committee and the IAEA. As a result the IAEA Transport Regulations are both stand-alone and a part of the UN Committee's Model Regulations.

In 1977, in recognition of the rapid scientific and technological developments in the field of nuclear transport, the IAEA established a Standing Advisory Group on the Safe Transport of Radioactive Materials (SAGSTRAM). It reviewed the existing guidelines; established a system for future reviews; developed general guidelines and methods for establishing international coordinated research programmes and identifying the effects of these on the comprehensive revision of the regulations; designed an information collection and retrieval system for the worldwide volume of nuclear traffic; and considered recommendations for the further development of the transport regulations. Together the IAEA's Transport Regulations and supporting Safety Guides serve as the basis for the regulation of nuclear transport involving all international organizations and states with significant nuclear transport activities.

Unlike other aspects of the international governance of nuclear energy, the transport domain does not have its own international treaty dealing with safety.[17] Rather, the UN Model Regulations and therefore the IAEA's Transport Regulations are implemented through incorporation into various related international instruments. For air transport these have become mandatory through International Civil Aviation Organization (ICAO) Technical Instructions annexed to the 1944 Chicago Convention on International Civil Aviation. In addition the International Air Transport Association (IATA) has made the Model Regulations a prerequisite for the transport of dangerous goods by air. For sea transport the International Maritime Dangerous Goods Code has been made mandatory through incorporation into the 1974 International Convention for the Safety of Life at Sea (the SOLAS Convention). For instance, in 1997 the International Maritime Organization (IMO) incorporated the Code for the Safe Carriage of Irradiated Nuclear Fuel, Plutonium and High-Level Radioactive Wastes In Flasks on Board Ships (INF Code) into the Convention (Goldblat 2002: 113). For land transport, while there is no single international agreement which includes the UN Model Regulations, they are incorporated into such agreements as the Model Regulations of the United Nations Economic Commission for Europe and the Regulations Concerning the International Carriage of Dangerous Goods by Rail. Even states that are not party to these agreements may decide and are encouraged to use the regulations as a basis for national legislation.

For decades nuclear shipments have taken place worldwide largely without serious incident and unnoticed by the general public. With reference to the United States, which has the largest and most dispersed civilian nuclear power programme, 'The government and the nuclear industry have been transporting nuclear materials, including a modest amount of commercial spent fuel, for decades, without incident' (Smith 2006: 274). There are two exceptions to this low profile. One is plutonium shipments from France to Japan which pass through Southeast Asia, including pirate-infested waters such as the Malacca Straits, and which require military escort. Greenpeace and others have protested these shipments in part because they oppose a plutonium-based fuel cycle but also because of safety and security concerns. A second issue, arising from opposition to nuclear power but also due to safety and security concerns, has been the movement of nuclear material within Europe, particularly in the UK and Germany. In the United States there has been continuing controversy over the robustness of transportation casks for spent fuel and high-level nuclear waste (Smith 2006: 274–275).

Nuclear emergency preparedness and response

Although triggered only after an aspect of nuclear safety has failed, emergency preparedness and response systems are vital in convincing the public that any nuclear accident can be appropriately handled. Global governance in this area consists of two key conventions and several mechanisms, all of which have emerged as a result of Chernobyl, with the IAEA playing its usual central role.

The Convention on Early Notification of a Nuclear Accident (CENNA) was adopted by a special session of the IAEA General Conference in September 1986 and entered into force in October 1986. As of October 2009 there were 106 contracting parties (states and international organizations) and 70 signatories (IAEA 2009b).[18] The Convention applies when an accident has the potential to, or results in, the release of radioactive material across international borders with consequences for the safety of another state. Unlike the CNS, it covers, in addition to nuclear reactors, other types of facilities such as nuclear fuel cycle facilities, radioactive waste management facilities, nuclear fuels or radioactive waste in transport or storage and radioisotopes. In the event of a nuclear accident on its territory a state party must provide full details to the IAEA and any state which is or may be physically affected. The IAEA is mandated in turn to inform all IAEA members, relevant international intergovernmental organizations, or any other states which may be affected. To achieve rapid notification each state party is obliged to ensure that the IAEA and other state parties are aware of the competent national authorities and an emergency point of contact.

The Convention on Assistance in the Case of a Nuclear Accident or Radiological Emergency (CACNARE) was adopted by the same special session of the IAEA General Conference in September 1986 and entered into force in October 1986. As of October 2009 there were 104 contracting parties (states and international organizations) and 68 signatories (IAEA 2009a).[19] In the event of a nuclear accident a state party may call on any other state party or international organization for assistance.

The IAEA was given a central role by CACNARE in making resources available, liaising with provider states and, if requested, coordinating assistance at the international level, including cooperating with other international organizations.

Over the longer term the IAEA is tasked with collecting information on experts, equipment and materials available to assist in emergencies and disseminating methodologies, techniques and research into response techniques. The IAEA is also tasked with developing training programmes, transmitting requests for assistance, establishing radiation monitoring programmes, conducting feasibility studies of monitoring systems, and promulgating safety standards dealing with emergency preparedness and response.

The Agency has taken specific steps to fulfil these requirements. At a state's request it dispatches Emergency Preparedness Review Teams (EPREV) to evaluate emergency preparedness and recommend improvements. To establish coordination in advance it has convened biennial meetings of the 'competent' authorities that state parties of the two conventions are meant to designate to handle nuclear accidents. In 2003 a National Competent Authorities' Co-ordinating Group (NCACG) was established. An Incident and Emergency Centre was established in 2005 to coordinate the provision of assistance and permit the effective sharing of information between states, their competent authorities, international organizations and technical experts.

The Emergency Response Network Manual (ERNM) and the Response Assistance Network (RANET) are designed to improve coordination of assistance and promote emergency preparedness in member states. Unfortunately, by the end of 2008 only 14 member states had registered expert capabilities with RANET, which is insufficient if it is to become a global repository of information on national assistance offerings (IAEA 2009i: 10).

The IAEA Response Plan for Incidents and Emergencies (REPLIE) details how the agency staff will organize in response to an emergency. At the request of the IAEA General Conference, the IAEA Secretariat is currently developing a unified system that will replace the current Early Notification and Assistance Conventions (ENAC) website and the Nuclear Events Web-based System (NEWS) (IAEA 2009i: 11).

An Inter-Agency Committee on Response to Nuclear Accidents (IACRNA) with around 20 members has been established to coordinate the response of relevant international organizations in the event of a nuclear accident.[20] IACRNA has developed a Joint Radiation Emergency Management Plan of the International Organizations (JREMPIO). The JREMPIO describes the roles and responsibilities of the various organizations, sets out the interfaces among them and with states, and establishes a framework for emergency preparedness. Emergency response exercises, coordinated by ICARNA and the IAEA, have also been held to increase preparedness. A most recent example is ConvEx-3 (2008) at the Laguna Verde reactor in Mexico. Seventy-five countries and nine international organizations participated (IAEA 2009i: 11). Work on a Code of Conduct on International Emergency Management (IAEA 2006b: 5) has, however, floundered, an example of the difficulties sometimes facing attempts to strengthen global governance in the nuclear safety area.

Since the Chernobyl disaster the global governance of emergency preparedness and response has gone from being non-existent to the complicated system of treaties, arrangements and measures described above. This area is probably the most striking example of the tendency of global nuclear governance, especially in the hands of the IAEA, to spawn multiple, seemingly ad hoc and uncoordinated initiatives in response to particular needs. It is hard to imagine newcomers to nuclear energy being able to easily absorb the intricacies of this system, which cries out for rationalization. It is also difficult to fully assess the system's adequacy until it is tested during an actual disaster, which is then too late. According to the NEA the systems established by the two treaties are becoming outdated and need revision. The IAEA's Nuclear Safety Review for 2008 notes bluntly that while states with nuclear installations tend to have adequate preparedness and response capabilities to deal with local incidents and emergencies, only a few have the capacity for responding to a major nuclear emergency (IAEA 2009i: 7). If and when increasing numbers of states acquire nuclear power plants they will need to be drawn tightly into this system, beginning with ratification of the two major conventions and followed by full compliance with their obligations.

The international nuclear accident liability regime

The international legal regime governing nuclear liability is the oldest, least understood and most fragmented aspect of global nuclear energy governance. It also has the lowest levels of state participation. The regime emerged in the early 1960s, which makes the original liability conventions the first multilateral treaties governing any aspect of nuclear power generation. They were seen at the time as vital in enticing power companies to invest in an unfamiliar, potentially dangerous form of electricity generation. The regime has been a continuous work-in-progress ever since.

The liability regime is an important part of global nuclear safety since it is the only dedicated legal mechanism by which an operator can be held internationally accountable for a nuclear accident that causes trans-boundary deaths, injuries and damage. It aims to sustain public confidence in nuclear energy by ensuring that those harmed are adequately compensated no matter where the accident occurs or whose fault it is. By imposing responsibility on operators, the regime reinforces the incentives for them to run their nuclear reactors safely. Yet in limiting the liability of the operator, with the state and the international community guaranteeing to fund damages above the limited amount, the regime also reinforces the interest of the state in properly regulating the industry. Moreover, it encourages financiers and utilities to invest in nuclear energy, encourages insurance companies to insure operators for limited liability and lowers operators' insurance premiums. Finally, it provides a stronger legal protection against unlimited liability for vendors that operate outside their own countries (MacLachlan 2008e: 6).

But the regime has paradoxical, less welcome aspects. First, it creates moral hazard for the nuclear industry and violates the 'polluter pays' principle. By limiting liability and thereby reducing the industry's incentives for relentlessly pursuing nuclear safety the regime attenuates the much vaunted principle that the operator is ultimately responsible for the safety of its nuclear reactor. This is particularly of concern when cost pressures in a deregulated market push in the direction of cutting expenses. Knowing that the government will foot the bill for a catastrophic accident (and will assume part of the blame due to the perceived failings of the regulator), the industry may be less driven in pursuing safety than it might otherwise be.

A second problematic aspect is that the system provides a small but hidden subsidy to the nuclear industry (no other energy sources have their own dedicated trans-boundary liability regime).[21] It renders nuclear energy cheaper than it would be if the full costs of private liability insurance were internalized in the price of nuclear electricity and thereby privileges it over other forms of energy generation.[22] A third difficulty is the regime's complexity and fragmentation. Even IAEA Director General Mohamed ElBaradei once remarked that, 'the provisions of the liability conventions, and the relationships between them, are not simple to understand' (ElBaradei 2007). The European Commission has begun a study aimed at 'simplifying' the EU regime which could result in operator liability pooling similar to the US scheme (MacLachlan 2008e: 7). In 2007 the IAEA published a comprehensive study and authoritative interpretation of the Vienna nuclear liability regime prepared with the assistance of an International Expert Group on Nuclear Liability (INLEX) appointed by the Director General (IAEA 2007e).

The main element of complexity comes from the fact that the regime is based on two separate international legal frameworks that the international community has constantly tampered with and attempted to cobble together. Each framework encompasses more than one international treaty and several amendments and additions that add further rights and obligations. The oldest by a few years is the Paris/Brussels framework, established under the auspices of the OECD/NEA, covering most OECD member states. The Vienna framework, under IAEA auspices, was intended to be universal. An attempt to modernize the regime came with the 1997 Protocol to Amend the Vienna Convention and the 1997 Convention on Supplementary Compensation for Nuclear Damage. The latter, in particular, purportedly 'provides the framework for establishing a global regime with widespread adherence by nuclear and non-nuclear States' (IAEA 2007e: 1). Despite their complexity, the main principles and content of the nuclear liability conventions are today internationally accepted as appropriate legal means for dealing with nuclear risks. They form, for instance, the international yardstick for assessing whether national nuclear liability legislation is adequate: the IAEA enjoins legislators to consider aligning their domestic legislation with the conventions (IAEA 2003b: 108).

In addition there is a treaty dealing specifically with maritime nuclear accidents, the 1971 Convention Relating to Civil Liability in the Field of Maritime Carriage of Nuclear Material. It entered into force in 1975 and its depositary is the International Maritime Organization (IMO). It limits the liability of nuclear operators that transport nuclear material. As of November 2009 it had only 17 state parties.

Table 4.1 Nuclear liability regime: signatories, ratifications and entry into force

	Signatories	*Parties*	*In force*
Paris/Brussels Framework			
1960 Paris Convention on Third Party Liability in the Field of Nuclear Energy	18	16	April 1968
1964 Additional Protocol	18	16	April 1968
1982 Protocol	18	16	October 1988
2004 Protocol	1	1	No
1963 Brussels Supplementary Convention	15	12	December 1974
1964 Protocol	15	12	December 1974
1982 Protocol	15	12	January 1988
2004 Protocol	2	2	No
Vienna Framework			
1963 Vienna Convention on Civil Liability for Nuclear Damage	36	14	November 1977
1997 Protocol	5	15	October 2003
1988 Joint Protocol Relating to the Application of the Vienna Convention and the Paris Convention	26	22	April 1992
1997 Convention on Supplementary Compensation for Nuclear Damage	13	4 Contracting States	No

Sources: 1960 Paris Convention latest status as of 10 June 2009 www.nea.fr/law/paris-convention-ratification.html; 1963 Brussels Supplementary Convention latest status as of 10 June 2009 www.nea.fr/law/brussels-convention-ratification.html;1963 Convention on Civil Liability for Nuclear Damage last change of status December 2008 www.iaea.org/Publications/Documents/Conventions/liability.html; 1997 Protocol last change of status July 2003 www.iaea.org/Publications/Documents/Infcircs/1998/infcirc566.shtml; 1998 Joint Protocol Relating to the Application of the Vienna Convention and the Paris Convention last change of status July 2009 www.iaea.org/Publications/Documents/Infcircs/Others/inf402.shtml; 1997 Convention on Supplementary Compensation for Nuclear Damage last change of status May 2008 www.iaea.org/Publications/Documents/Infcircs/1998/infcirc567.shtml.

However, unlike most other areas of nuclear global governance, states have proved remarkably reluctant to become parties to the liability conventions and protocols. Fewer than half the world's nuclear power plants are currently covered by the regime (MacLachlan 2008e: 6). Tellingly, states with the largest civil nuclear programmes have stayed away, which is 'clearly a disincentive for other states to join' (Rautenbach *et al.* 2006: 34). Three of the legal instruments have attracted so few parties that they have not yet come into force. In the case of the 2004 protocols to the Paris and Brussels Conventions there is a risk that they will be overtaken by a 2004 EU Environmental Liability Directive that exempted nuclear damage on the grounds that it would be covered by the international conventions, but which provides for a review in 2014 'to see if it is still justified' (MacLachlan 2008e: 7). US ratification of the Convention on Supplementary Compensation in May 2008 was expected to be a major fillip, but so far even this has not induced other states, notably nuclear energy producers, to join. INLEX is attempting to increase adherence by holding regional workshops.[23]

One of the reasons given for the lack of adherence is dissatisfaction with various provisions of the conventions. Many states view the minimum liability provisions as too high. Others find the broadened definition of nuclear damage that includes environmental damage, or the extended geographical scope, to be unpalatable (Schwartz 2006: 49). Still others see the preferential treatment given to extra-territorial victims as discriminatory (Schwartz 2006: 52). Differences between the Vienna Convention and the Brussels Supplementary Convention have led to claims by some states that that it would be 'hard to envision signing two complementary conventions with different mechanisms, allocation rules and beneficiaries' (Dussart 2005: 24). Some national practices differ wildly from the model reflected in the international regime. Germany, for instance, insists on unlimited operator liability (MacLachlan 2008e: 7).

The low levels of participation by states are highly problematic for the regime since it needs adherence by as many countries as possible to create a sizeable compensation fund. Even so, the maximum amounts foreseen in the revised Paris and Brussels Conventions, totalling €1.5 billion, although much more than currently provided, would not come close to covering the costs of a catastrophic accident. Many billions of dollars in compensation have been paid out in the former Soviet Union and some Western European states for damage from the Chernobyl accident and the payments continue because land is still contaminated and long-term health effects are being claimed. According to Ann MacLachlan:

> A threat hanging over all the liability regimes is that as the coverage has been broadened, in the light of experience from Chernobyl, to include more types of damage, longer claim periods, and higher compensation, the insurance industry has baulked at providing the coverage operators must legally contract under national liability legislation.
>
> (MacLachlan 2008e: 7)[24]

Since the terrorist attacks of 11 September 2001 the insurance industry has looked much more critically at its exposure to the risk of unlikely but high consequence events.

Conclusions

The global nuclear safety regime has markedly improved since the wake-up call of the 1986 Chernobyl accident and its dramatic demonstration of radioactive cross-boundary effects. Old Chernobyl-style reactors have been closed and other Soviet types retrofitted for better safety; international conventions have been negotiated; and international standards clarified and promoted. Industry itself has become more safety conscious, aware that a major nuclear accident anywhere is a major accident everywhere and could kill prospects for a revival. Peer review, via the IAEA and WANO, is an innovation that appears to work well, making up for the lack of monitoring and verification. Still, alarming incidents continue to occur even in a well-regulated industry like that of the United States. A lack of transparency prevents outsiders from knowing the true state of many countries' civil nuclear installations. The civilian nuclear fuel cycle beyond reactors themselves is largely ungoverned by the global governance regime and needs urgent attention (as do research reactors). The governance of spent fuel and nuclear waste has seen improvement with negotiation and implementation of the Joint Convention but no state can claim to have dealt with the issue of long-term high-level radioactive waste satisfactorily. The emergency preparedness conventions and the associated complex (albeit confusing) web of supportive measures are an advance on what existed before Chernobyl, but most states, especially those in the developing world, do not have in place the necessary capacities to deal with a major nuclear emergency. The nuclear liability arrangements, meanwhile, are seriously under-subscribed and under-funded, as well as exhibiting the type of complexity and inflexibility that is likely to deter new entrants.

The current global governance regime for nuclear safety is thus complex, sprawling and based on a variety of treaties and implementation mechanisms that have arisen in different eras to meet particular needs. It does, however, now seem to have all of the necessary components in place, with the exception of legally-binding safety agreements for fuel cycle facilities (and research reactors). To cope with increased use of nuclear energy it does not need wholesale reform or major additions but rather: universal adherence to existing treaties; enhancement and rationalization of existing mechanisms; and increased human and financial resources, including for regulatory purposes. Global and national nuclear safety need to be a permanent work-in-progress and complacency and regression avoided.

5 The current status of global nuclear governance
Nuclear security and non-proliferation

Nuclear security and the non-proliferation of nuclear weapons are increasingly seen as closely linked. Originally nuclear security was designed to prevent unauthorized access to nuclear materials that may have been harmful to individuals and the general public, and to prevent sabotage whether by outsiders or insiders. A dawning realization that terrorists may be able to fashion a crude nuclear device from purloined fissile material,[1] combined with fears of a terrorist attack on a nuclear facility that could replicate the radioactive effects of a nuclear weapon, has led to increasing convergence of the nuclear security and non-proliferation elements of global governance. This chapter thus treats them together, mindful that several of the global institutions, treaties and arrangements considered below have relevance to both security and non-proliferation.

The international nuclear security regime

Security affects the nuclear industry in a way that it does not affect other forms of energy generation. This is partly a legacy of the highly secretive nuclear weapons programmes from which civilian applications of nuclear energy emerged. It is also due to the strategic nature of the facilities and nuclear materials involved. Large nuclear power plants or other facilities may make tempting targets for saboteurs, while nuclear materials may be purloined for use in nuclear weapons or radiological weapons (also known as radiological dispersal devices or RDDs). Hence nuclear security is considered the exclusive preserve of sovereign states in a way that nuclear safety is not, making global governance in this area much more challenging. Since nuclear security and radiological protection measures necessarily involve key national functions such as law enforcement and control over access to information, states are 'understandably reluctant to expose their sovereign security and law enforcement practices to external scrutiny, let alone anything resembling external regulation' (IAEA 2003a: 145). Moreover, as Matthew Bunn points out, 'any test or assessment that revealed particularly urgent vulnerabilities would be especially closely held' (Bunn 2009: 115). As the IAEA judiciously puts it: 'the responsibility for nuclear security rests entirely with individual States' (IAEA 2006d: 1).

Anther reason for the contrast in global governance in the safety and security domains is that while safety has been amenable to quantitative probabilistic risk assessment, security threats are much more difficult to quantify because of 'intelligent adversaries' and the paucity of data due to the few attacks on nuclear facilities that have occurred (Ferguson and Reed 2009: 59). Another major difference is that 'safety culture has evolved to become more open about admitting mistakes in a "no fault" environment that should work to correct mistakes without seeking retribution on workers who have made mistakes or whistleblowers' (Ferguson and Reed 2009: 59): the nuclear security area by contrast is characterized by secrecy and a lack of public accountability.

What Chernobyl did for nuclear safety, 11 September 2001 has done for nuclear security. The audacity of the international conspiracy that led to the attacks of 11 September 2001 on the continental United States has heightened awareness about two particular threats: the potentially catastrophic effects of a terrorist attack on a nuclear reactor or other nuclear facility, in effect using it as a radiological weapon; and, second, the possibility that a well organized and well funded group like Al Qaeda might seize nuclear material from the civilian nuclear fuel cycle for a nuclear weapon or RDD and might actually be able to use it for that purpose. Since 11 September 2001 there has been laudable action to strengthen the hitherto patchy international nuclear security regime for civilian nuclear energy.

Nonetheless, the international nuclear security regime remains nowhere near as extensive, advanced or entrenched as the regime for nuclear safety. There are fewer treaties, a less widely accepted set of recommended security principles and practices, little collaboration between nuclear plant operators worldwide, as in the case of WANO for nuclear safety, practically no peer review and an abiding sense that nuclear security is too sensitive an issue to be subject to global governance. Russia is reportedly especially opposed to international 'interference' in national nuclear safety (ICNND 2009: 117). As Roger Howsley puts it, 'The pervasive secrecy surrounding nuclear security means that no global mechanism is in place to identify the worst security performers and help them come up to the level of the best performers' (Howsley 2009: 204). Ferguson and Reed note that:

> While improvements in nuclear safety have built on more than 50 years of experience in the commercial nuclear industry, the standards of excellence emulated from other nuclear organizations, and the decades long experience of the IAEA in developing nuclear safety standards, nuclear security has not received as much attention and resources from the communitarian perspective.
>
> (Ferguson and Reed 2009: 59)

This section considers the international nuclear security regime, outlining the existing treaties, mechanisms and assistance measures currently in place. As in other areas, a key role is played by the IAEA, but other organizations are also involved. While the regime is less extensive than that for nuclear safety, there is some overlap between the two that provides a degree of mutual reinforcement.

Convention on the Physical Protection of Nuclear Material

The 1980 Convention on the Physical Protection of Nuclear Material (CPPNM) was opened for signature in March 1980 and entered into force in 1987. As of January 2010 there were 142 state parties (as well as Euratom) and 45 signatories (IAEA 2010d). It is the only legally binding multilateral treaty relating to the physical protection of nuclear material.

The purpose of the CPPNM is to commit states to ensure that nuclear material for civilian purposes under their jurisdiction is protected during international transport. It does this in three ways. First, it establishes legally prescribed protective levels for nuclear material during such transport. Second, it seeks criminalization by states of the theft of nuclear material. Third, it promotes international cooperation in prosecuting offences and responding in the event of a breach. The treaty does not apply to nuclear material for military purposes or radioactive sources.

The CPPNM contemplates two protection scenarios: material stored in preparation for or immediately after international transport; and during international transport itself. It outlines requirements for control of access to nuclear materials, including: surveillance by guards or electronic devices; physical barriers; limited and controlled points of entry; and close communication between surveillance personnel and response forces. Parties are obliged not to permit export of nuclear material unless assured that it will be protected during transport at the prescribed levels. Parties also must not import nuclear material from a non-state party unless assured that the material will be protected during transport at the levels provided for in the convention. Additionally, parties are required not to allow the transit of nuclear material through their own territory unless it is so protected. Each party must identify to all other parties, either directly or through the IAEA, a central national point of contact with responsibility for physical protection of nuclear material and for coordinating recovery and response operations in the event of a breach.

While the treaty contains provisions for review conferences every five years, these are aimed at assessing the implementation of the convention as a whole, not the compliance of individual parties. There is no peer review mechanism, as in the case of the CNS, nor does the IAEA have any particular role beyond transmitting information about national contact points. Monitoring or verification of compliance is completely absent. There is the usual dispute resolution mechanism, involving referrals to the International Court of Justice (ICJ), but these relate to interpretation of the treaty, not non-compliance. However, the IAEA does provide states, on request, with advisory, review and other services to help them, among other things, assess and improve their compliance with the CPPNM.

Amendment to the Convention on the Physical Protection of Nuclear Material

In 1998, a group of experts convened by the IAEA Director General recommended that consideration be given to revising the CPPNM to extend it to domestic use, storage and transport. Negotiations on an amendment stretched over many years, but were formally concluded at a diplomatic conference in Vienna in July 2005. The endgame of the negotiations was stimulated by the nuclear 'near miss' that some considered the events of 11 September 2001 to have been. The Amendment created a legally binding regime requiring each state party to the CPPNM to establish and maintain an 'appropriate physical protection regime' for nuclear material in use, storage and transport and for nuclear facilities anywhere under its jurisdiction. Such a national regime should be designed to prevent theft, establish a rapid response capacity to locate and recover missing or stolen nuclear material, protect against sabotage of nuclear material or nuclear facilities, and mitigate the consequences of any successful sabotage. Each party must embed the treaty in its legal system, establish a legislative and regulatory framework to govern physical protection, and designate a competent authority responsible for domestic implementation and a point of contact which should be imparted to all other parties and the IAEA.

The Amendment to the CPPNM was adopted in July 2005. As of January 2010 there were 33 parties (IAEA 2010d). The Amendment is not yet in force, as this is contingent on ratification by two-thirds of the original 112 state parties to the CPPNM. One reason why early entry into force is so desirable is that the IAEA can then begin directing its advisory and expert services towards ensuring compliance by states with nuclear safety standards domestically as well as during international transport. This would amount to a major strengthening of the global nuclear security regime.

International convention for the suppression of acts of nuclear terrorism

Recognition of the threat of nuclear terrorism clearly preceded 11 September 2001, derived in large part from the recognized risk that terrorists or other unauthorized persons could obtain 'loosely' secured fissionable material from the former Soviet nuclear programmes (both peaceful and military). It was Russia, therefore, which took the initiative in the United Nations General Assembly to propose an international instrument on nuclear terrorism. An ad hoc committee was accordingly established by the General Assembly in 1996 to begin discussions on conventions banning 'terrorist bombings and nuclear terrorism'. In 1998, the Committee began negotiations on an International Convention for the Suppression of Acts of Nuclear Terrorism (ICSANT), based on a text proposed by Russia. The events of 11 September 2001 provided a stark rationale and final impetus for wrapping up the drawn-out negotiations. The Convention was adopted by the General Assembly in April 2005, opened for signature in September 2005 and entered into force in July 2007. As of January 2010, there were 63 state parties and 115 signatories (UN 2005).

ICSANT establishes a wide variety of offences in relation to nuclear terrorism. Each party is obliged to establish the offences within its domestic criminal law, ensuring that the penalties take into account the grave nature of nuclear terrorism. ICSANT also obliges parties to exchange information on nuclear terrorism threats and assist each other in criminal proceedings. ICSANT applies to all nuclear materials and facilities, including those used in civilian nuclear power programmes.

In terms of international governance, although the treaty names the UN Secretary-General rather than the IAEA Director General as depositary and therefore it is not considered to be within the IAEA's 'family' of treaties, the IAEA does assume several important treaty functions. If a state seizes control of any radioactive material, devices or facilities following the commission of an offence, that party must ensure that they are placed under IAEA safeguards and must 'have regard' for IAEA physical protection recommendations and health and safety standards. In doing so, the state party may call on the assistance of the IAEA. In addition, a state party that seizes material, a device or a facility is obliged to inform the IAEA Director General of the manner in which such an item was disposed of or retained.

While a valuable addition to the nuclear security regime, ICSANT does not have any monitoring, verification or compliance provisions or system of peer review or accountability. Nor does the convention have provision for review meetings, but simply enjoins the parties to consult one another on its implementation. Amendments may be approved by a specially convened meeting of parties. Clearly ICSANT warrants strengthening to ensure its implementation and make its parties more accountable and transparent in terms of their compliance with its obligations.

African Nuclear Weapon-Free Zone Treaty

Surprisingly, given their usual exclusive focus on the regional non-proliferation of nuclear weapons, there is one nuclear weapon-free zone treaty, the 1986 African Nuclear Weapon-Free Zone Treaty (ANWFZ), also known as the Treaty of Pelindaba, that contains provisions for ensuring the physical security of nuclear materials.[2] An initiative of South Africa, the Treaty was opened for signature in April 1996 and entered into force with its twenty-eighth ratification on 15 July 2009 (Broodryk and Stott 2009). Geographically it covers the entire African continent. Under Article 10 of the treaty, state parties are legally obliged to maintain the 'highest standards of security and effective physical protection' of nuclear materials, facilities and equipment. Each party undertakes to apply measures of physical protection equivalent to those provided for in the CPPNM and IAEA security guidelines. The treaty also bans attacks on nuclear facilities, again the only NWFZ to contain this provision.

To facilitate exchanges of information, consultations and, ultimately, compliance, the African Commission on Nuclear Energy (AFCONE) is supposed to be established after entry into force, but preparations have only just begun to achieve this. On receiving a detailed inspection report from the IAEA indicating a breach, AFCONE is supposed to meet in extraordinary session and make recommendations to the party concerned and to the African Union (AU). If necessary, the AU may refer the matter to the United Nations Security Council. This is the most explicit compliance language applicable to nuclear security in any multilateral treaty and could prove useful if African states, apart from South Africa, succeed in acquiring civilian nuclear energy. The member states of other NWFZs could be encouraged to emulate the African zone in its attention to nuclear security.

Security Council Resolution 1540

Adopted in April 2004 by the United Nations Security Council under Chapter VII of the UN Charter, which makes it legally binding, Resolution 1540 obliges all states to refrain from providing support or assistance to non-state actors seeking to acquire so-called weapons of mass destruction (WMD) – normally taken to mean nuclear and radiological, as well as chemical and biological, weapons. The resolution[3] also requires states to adopt and enforce appropriate and effective laws that prevent non-state actors acquiring WMD or related materials and technologies. With respect to nuclear material, 1540 requires all states to develop and maintain: measures to account for and secure such items; 'appropriate and effective' physical protection measures; 'appropriate and effective' border controls and law enforcement agencies; and national export and trans-shipment controls. The resolution is thus both a nuclear security and a non-proliferation measure. Unfortunately, the Council did not prescribe the characteristics of the steps that states were required to take, nor did it define 'appropriate' or 'effective'. All UN member states are required, however, to report to the Council on their compliance with the resolution. To ensure implementation and facilitate compliance, the resolution established a 1540 Committee comprising Security Council members, furnished belatedly with some secretariat and expert assistance.

Resolution 1540 is a valuable, novel addition to global governance in the nuclear area, especially in drawing attention to the importance of national implementation measures to ensure nuclear security and in obliging all states, not just those party to the relevant international treaties, to comply with the will of the Security Council in this regard. It can also play a useful role in matching needy states with those which can provide assistance.

Yet it would be wise not to put too much store in the resolution in dealing with the challenges of a nuclear energy revival. First, the resolution is very broad. It deals with all types of WMD and even within the nuclear category the focus is on legacy materials and installations from weapons programmes, not civilian uses. While civilian nuclear energy will benefit in the long run from better national nuclear security and non-proliferation measures, nuclear power generation barely rates as a concern of 1540. Second, compliance with 1540 has been slow and uneven. As of April 2010, 157 countries out of 192 (82 per cent), along with the EU, had submitted national reports to the 1540 Committee (1540 Committee 2010) but this level of compliance has taken years to achieve. Many developing countries lack the capacity and expertise to even meet the reporting requirements of 1540, much less undertake the substantive steps to strengthen national implementation required. Sub-Saharan African countries, for instance, are 'little inclined to divert scarce resources for implementing nonproliferation obligations' (Heupel 2007: 6–7). Among such states are several that have expressed interest in nuclear energy (Ghana, Kenya, Namibia, Nigeria, Senegal and Sudan).

A Stanley Foundation study confirms that implementation remains slow and uneven, 'in part due to the incredible diversity of different national circumstances and the lack of rationalized machinery at the global level' (Stanley Foundation 2009: 2). Encouragingly, regional and sub-regional organizations have begun playing an increasingly significant role in assisting in 1540 implementation (Scheinman 2008). Meanwhile the 1540 Committee has enlisted the IAEA's help in recommending better protection of nuclear facilities and materials from theft and sabotage (Bunn 2007), but technical assistance is available in any case directly from the Agency for member states that request it.

Role of the International Atomic Energy Agency

As in other nuclear matters, the IAEA plays a critical role in helping implement the existing legal instruments concerning nuclear security, as well as advising and assisting states in fulfilling their international and national obligations regarding physical protection for both nuclear materials and nuclear facilities. In 2007 a review of the IAEA's security programme chaired by Roger Howsley, currently inaugural director of the World Institute of Nuclear Security (WINS), concluded that 'the IAEA security team is doing a fantastic job' (Howsley 2009: 204). However, compared to its nuclear safety programme, the agency's nuclear security programme is relatively small and under-funded (Ferguson and Reed 2009: 59).

Since 1972 the IAEA has issued non-binding but authoritative recommendations on the physical protection of nuclear material and nuclear facilities. These are updated periodically, most recently in 1998 (IAEA 1998). In 2006 the agency launched its Nuclear Security Series of documents, to assist states in establishing a coherent nuclear security infrastructure. They are structured in the same way as its documents on nuclear safety, with a similar three-level schema (Fundamentals, Recommendations and Implementing Guides and Technical Guides) presumably in an effort to encourage states to treat them the same way.[4] International experts assist the IAEA Secretariat in drafting these publications. The drafting and review process takes account of confidentiality considerations and according to the agency 'recognizes that nuclear security is inseparably linked with general and specific national security concerns' (IAEA 2008g).

The IAEA offers an impressive array of assistance to states in the nuclear security arena, much of it grouped under its Three-Year Plan of Activities to

Protect against Nuclear Terrorism. The third plan, covering 2010–2013 was adopted by the IAEA General Conference in September 2009 (IAEA 2009k). According to the agency, the programme of three-year plans has achieved 'sufficient maturity to evaluate its own accomplishments and shortcomings, set meaningful priorities and indicators of success, and take into consideration the evaluations and inputs of other interested stakeholders and groups, including donors to the Nuclear Security Fund (NSF)' (IAEA 2008j: 1). In the last couple of years the Fund has dispersed around $15–16 million annually in various nuclear security projects. Funding comes from extra-budgetary donations by just a few states, in addition to 'in kind' contributions. A major new source of funding is the EU Strategy against Proliferation of Weapons of Mass Destruction. Unfortunately 90 per cent of the funds donated come with conditions, making 'setting overall programmatic priorities difficult' (IAEA 2008j: 2). While developing states have, laudably, taken advantage of the three-year programmes, the agency reports a low participation rate by developed countries, which may consider their nuclear security needs taken care of or may be sensitive about confidentiality or their sovereign prerogatives in this area.

There are a variety of nuclear security review services and missions furnished by the IAEA to states. The International Physical Protection Advisory Service (IPPAS) is the IAEA's primary mechanism for evaluating physical protection arrangements. It conducts detailed reviews of the legal and regulatory infrastructure of a requesting state, determines the level of compliance with the CPPNM and compares it with IAEA standards and international best practice. As of October 2009, 46 IPPAS missions had been completed in 31 states (including follow-up missions in ten states) in all regions (IAEA 2008h; Gregoric 2009). The IAEA provides follow-up assistance such as training, technical support and more targeted assessments. In addition, the Agency's International Nuclear Security Advisory Service (INSServ) conducts missions to assist in identifying a state's nuclear security requirements and how it can meet them. INSServ generates a report which can serve as the basis for cooperation between the state and the IAEA and for bilateral nuclear security assistance. The agency reports that 29 INSServ missions had been conducted to 30 June 2008 (IAEA 2008g). The International SSAC Advisory Service (ISSAS), meanwhile, provides requesting states with recommendations regarding improvements to their State System of Accounting and Control (SSAC), the basis, since 1993, of the IAEA's strengthened safeguards system. The service contributes to safety and security by ensuring that states can adequately account for their nuclear material. International Team of Experts (ITE) advisory missions may also be dispatched to an IAEA member state to inform national policy-makers about the need for adherence to the international legal framework governing nuclear material and how to implement it domestically.

In addition to these activities, the Agency is also attempting more holistic and business-like approaches to nuclear security capacity-building. Integrated Nuclear Security Support Plans (INSSP), based on findings from nuclear security support missions, attempt to provide states, in contrast to previous ad hoc measures, with a 'holistic' approach to nuclear security capacity building. The plan is individualized to meet the needs of each state. From June 2007 to July 2008 the agency developed six INSSPs in cooperation with state authorities in Brazil, China, Pakistan, Peru, Qatar and Saudi Arabia, bringing the total to 44 since the programme began (IAEA 2008j: 2, 18). The IAEA has also recently developed a conceptual approach for the establishment and maintenance of national Nuclear Security Support Centres to foster a 'systematic, business-oriented approach' to nuclear security (IAEA 2008j: 17). The centres will serve as a focal point for sustainable and continued access to knowledge, skills and capabilities. It appears that no centre has yet been established (IAEA 2009m).

The IAEA also has an increasingly important role in seeking to uncover illicit trafficking of nuclear materials. Although currently more concerned with 'legacy' nuclear materials that may still be leaking from the states of the former Soviet Union, the IAEA's role in seeking to uncover illicit trafficking is relevant to the future of civilian nuclear energy. As security is improved at Russian military nuclear sites traffickers may look to the civilian sector for opportunities.

Established in 1995, the IAEA's Illicit Trafficking Database (ITDB) is designed to facilitate the exchange among states of authoritative information on reported incidents of illicit trafficking in all types of nuclear materials and radioactive sources. The ITDB covers unauthorized acquisition (for example, theft), supply, possession, use, transfer and disposal of nuclear and other radioactive materials, whether intentional or unintentional, and whether or not international borders are crossed. The ITDB also covers unsuccessful or thwarted acts, accidental loss of materials and the discovery of uncontrolled materials. All types of nuclear materials (uranium, plutonium and thorium), all naturally occurring and artificially produced radioisotopes, and radioactively contaminated materials are included. No limit is placed on the amount of material that should be reported, its activity level or other technical characteristics. States are also encouraged to report scams in which non-radioactive materials are offered for sale as nuclear or radioactive materials. States are not obliged to contribute, since the database does not derive from a treaty obligation or other international agreement. The ITDB still mainly collects information from open sources, but seeks confirmation about its veracity from the member state concerned.

ITDB information is continuously analysed by the agency's staff to identify trends and patterns, assess threats, and evaluate weaknesses in material security and detection capabilities and practices (IAEA 2006a). The Secretariat produces quarterly and annual reports containing ITDB statistics and analysis. Participating states are also provided with regularly updated CD-ROM versions of the database. Communication with participating states is maintained through a network of national Points of Contact (POC). Meetings of the POCs are organized regularly to review the operation of the ITDB.

The Agency concludes that continuing incidents involving attempts to sell nuclear materials or radioactive sources indicate that there is a perceived demand for such materials on the illegal market. The majority of incidents have been supply-driven with no buyers. However, in some cases buyers and repeat offenders have been identified. Amateurishness and poor organization have characterized many trafficking cases, according to the IAEA, whereas well-organized, professional and demand-driven trafficking would be much more difficult to detect. Where information on motives is available, says the Agency, financial gain seems to be the principal motive. Some cases, however, showed an indication of malicious intent.

One of the difficulties with this reporting instrument, as with others in the nuclear safety and security area, is that not all states provide reports and not all provide the requisite information when they do report. Since July 2007 the IAEA has conducted eight regional information meetings for countries in Asia, Africa, the Middle East and Eastern Europe on illicit nuclear trafficking information management and coordination, in part to encourage participation in the ITDB but also to help strengthen national, regional and international capacities.

The IAEA has also formed the Border Monitoring Working Group (BMWG) to promote and coordinate multilateral and bilateral cooperation in establishing border monitoring capabilities. The IAEA itself assists states in establishing such capabilities. In 2008 alone it worked with 19 states, providing more than 260 items of equipment to improve detection and response capacities (IAEA 2008j: 13). The IAEA's Nuclear Security Equipment Laboratory (NSEL) helps ensure that border detection instruments meet technical and functional specifications.

The role of other organizations and initiatives

The Global Initiative to Combat Nuclear Terrorism, jointly proposed by Russia and the United States and launched in July 2006, aims to increase international cooperation in combating nuclear terrorism. Since its inception in 2006, the initiative has garnered the support of 76 countries for its statement of principles (US Department of State 2009c). As of June 2009 partners have hosted more than 30 workshops, conferences and exercises (US Department of State 2009a). The IAEA and EU participate as observers. While a useful addition to the expanding network of stakeholders in the nuclear security area, especially in bringing together the non-proliferation, counter-proliferation and counter-terrorism communities, the initiative has some drawbacks. It is far from universal, it has no standing institutional support, and until 2009 it involved only governments. However, at their June 2009 meeting members agreed to admit the International Criminal Police Organization (INTERPOL) as an observer and to promote participation by civil society and business (US Department of State 2009c). The impetus for the initiative is, moreover, the existence of terrorist threats involving nuclear weapons and existing nuclear materials, especially legacy materials from past weapons and civilian activities. It is thus not particularly attentive to the civilian nuclear power enterprise, although it could and should be extended to do so.

The World Institute of Nuclear Security (WINS) is a relatively new organization, launched in September 2008 and based appropriately in Vienna. It has been jointly initiated by two US-based non-governmental organizations, the Nuclear

Threat Initiative (NTI) and the Institute of Nuclear Materials Management (INMM), in cooperation with the US Department of Energy and the IAEA. It is being supported financially by NTI and governments, to date Norway and the United States. WINS is dedicated to helping secure all nuclear and radioactive materials globally so that they cannot be used for terrorist purposes. The organization promises to bring together nuclear security experts, the nuclear industry, governments and international organizations to focus on 'rapid and sustainable' improvement of security at nuclear facilities globally (WINS 2009). Currently, the size and nature of WINS' membership is unknown. WINS has inaugurated a newsletter for members and released its first publication, a *Best Practice Guide on Security Culture* (WINS 2009).

The impetus for the creation of WINS has yet again come from concern about existing stocks of nuclear weapons materials, especially legacy materials in the former Soviet states, and civilian uses of HEU and plutonium, such as in research and isotope production reactors. While its establishment fills a significant gap in the global governance regime for nuclear security, its relevance to civilian nuclear energy and hence to a nuclear revival remains to be seen. The task it has set itself with respect to existing weapons-grade materials is already daunting enough without tackling lesser security threats represented by an expansion of civilian nuclear energy. However, WINS' success in its chosen area of concern would pave the way for future attention to the security of the peaceful nuclear fuel cycle. Moreover, as it begins to instil observance of international security norms and expectations of a robust security culture in its member organizations, this may filter into all agencies concerned with nuclear security, including civilian plant operators.

The European Union (EU) Strategy against Proliferation of Weapons of Mass Destruction is partly devoted to nuclear security (IAEA 2008j: 18). The EU has pursued a series of so-called Joint Actions since 2004 in support of the IAEA's Nuclear Security Plans, providing substantial financial and other contributions. In focusing on Southeastern Europe, Central Asia, the Caucasus, North Africa and the Mediterranean Middle Eastern states, Africa and South-East Asia, the programme is covering many of the states currently seeking to acquire nuclear energy for the first time.

Assessing the effectiveness of the current nuclear security regime

Assessing the effectiveness of the current international nuclear security regime is problematic, especially in respect of national implementation of the regime's treaty-based obligations. There are no reporting or peer review requirements comparable to those found in either the Convention on Nuclear Safety or the Joint Convention on the Safety of Spent Fuel Management and on the Safety of Nuclear Waste Management. Nor is there effective peer review of domestic physical protection systems by a nuclear industry body, as in the case of the WANO for nuclear safety. As Roger Howsley notes, 'building a sense of urgency and commitment to nuclear security within the nuclear industry' is a challenge (Howsley 2009: 207). While employees in the industry 'are trained to focus on safety from the first day of their careers', the same is apparently not true of security.

While IAEA standards and advisory services are vital, the global implementation of the highest levels of nuclear security is limited by the voluntary nature of the standards themselves and of the parameters of the assistance provided. No current treaty provides the IAEA with the authority to insist on mandatory physical protection standards or other elements of nuclear security. Many states have long resisted such an approach. As in the nuclear safety area, there does, however, appear to be increasing acceptance that the IAEA's standards are the international benchmark against which performance should be measured. In his background report for the work of the 2008 Commission of Eminent Persons on the Future of the Agency, then Director General ElBaradei suggested that by 2020 'many of the nuclear security documents will have become, de facto or de jure, international security standards and incorporated into national security policies and regulations' (IAEA 2008a: 19). While this may happen, in the meantime states do not necessarily feel compelled to abide by them and some continue to treat them as merely recommendatory.

There is much less interest by states in taking advantage of the IAEA's nuclear security reviews compared to the safety realm, where demand threatens to overwhelm the agency's capacity. Moreover, the nuclear security establishment, for partly understandable reasons, exhibits an even greater lack of openness and transparency than the nuclear safety community. Roger Howsley argues, however, that nuclear security does not have to be as closed a topic as commonly imagined (Howsley 2009: 207). He records that when he worked for British Nuclear Fuels (BNFL) it conducted a national stakeholder dialogue from 1999 to 2005 that addressed many aspects of its operations, including nuclear reactors, though a nuclear security working group. The group, which included those with anti-nuclear views, was able to reach a consensus on around 60 recommendations, some of which BNFL adopted, improving security as a result. Howsley concludes that 'Despite the profoundly different positions held by members of the group, this provided clear evidence that properly facilitated meetings can be very productive and need not compromise security in any way' (Howsley 2009: 207). Canada's experience with wide-ranging community consultation to determine its nuclear waste policy, which also included security as one of its key objectives, was similar (NWMO 2005: 194–201).

Internationally, Security Council Resolution 1540 and its reporting requirements, the only existing mechanism that provides any public record of implementation efforts in the nuclear security arena, are a step in the right direction. But reporting is only at a general level and does not focus on nuclear security specifically, much less the narrower area of the security of nuclear electricity generation and its associated fuel cycle activities. Moreover, the limited role currently being played by the IAEA in implementation of Resolution 1540 does not adequately capitalize on the agency's strengths, resources and expertise, especially given the resource and capacity constraints faced by the 1540 Committee itself. ICSANT is another useful initiative in increasing awareness of the importance of nuclear security but it singularly fails to hold states to account in respect of implementation and compliance. The African Nuclear Weapon-Free Zone provisions on nuclear security are a useful precedent for other zones, but it is doubtful that they will be seriously implemented and monitored any time soon.

The most urgent problem facing the international nuclear security regime is that the Amendment to the CPPNM – the instrument that would create binding legal obligations to protect nuclear material in domestic use – has not drawn wide support, and consequently has not entered into force. When it does, the IAEA can at last begin to assess states' domestic nuclear security arrangements in the framework of compliance with the Amendment, which will go a long way towards strengthening the regime's current arrangements. As in the case of nuclear safety, nuclear security is a perpetual work in progress that warrants further attention in the light of the putative nuclear energy revival.

The global nuclear non-proliferation regime

Compared to the global governance regimes for nuclear safety and security, the non-proliferation regime is for the most part legally binding, relies heavily on monitoring and verification of compliance and has the most extensive range of treaties, institutions, mechanisms and informal arrangements. Due to the fact that it concerns vital national security interests, the regime is also the most prone to non-compliance controversies that periodically roil the waters of international peace and security and end up on the agenda of the United Nations Security Council. The NPT and the IAEA are the core components of the regime, but these are buttressed by several other important mechanisms ranging from multilateral treaties to unofficial 'coalitions of the willing', as well as informal norms, rules and principles.

The Nuclear Non-proliferation Treaty

Apart from the Statute that established the IAEA, the Nuclear Non-proliferation Treaty is the founding international legal instrument of the nuclear non-proliferation regime. It was opened for signature in 1968 and entered into force in 1970. It contains a set of bargains. In return for assistance in the peaceful uses of nuclear technology the non-nuclear weapon states (NNWS) agreed not seek to acquire nuclear weapons. Compliance would be verified by the IAEA by nuclear 'safeguards' and be subject to consequences in case of non-compliance, ultimately through referral to the UN Security Council. The NPT also prohibited the five designated existing nuclear weapon states (NWS) – China, France, the Soviet Union, the UK and the United States – from assisting NNWS to acquire nuclear weapons. In what is the weakest part of the treaty Article VI required 'negotiations in good faith' by all NPT parties (but by implication especially the NWS) to achieve nuclear disarmament.

Over the decades the NPT has proved its worth, helping avoid the world of 20-plus nuclear weapon states predicted in the 1960s, and gradually attracting parties so that today it is almost universal, albeit with three significant 'holdouts' – India, Israel and Pakistan – and one withdrawal, that of North Korea. In 1995, the treaty was extended indefinitely. Despite periodic warnings of its imminent demise it has endured, essentially because of the security benefits it confers on its members (although these seem to be under constant debate).

However serious cases of non-compliance – by Iraq, North Korea, Libya and Iran – have undermined confidence in the treaty. Just as insidious has been growing dissatisfaction with the NPT's arbitrary and apparently permanent concretization of two classes of states: those that had detonated a nuclear device before 1 January 1967 – which also happened to be the permanent members of the Security Council – and those which had not. Over the years the lack of progress towards complete nuclear disarmament (despite significant cuts in nuclear weapons since the Cold War ended), and the lack of accountability of the NWS in meeting their Article VI obligations, has increasingly put the NPT and the IAEA under strain. Attempts to constantly strengthen nuclear safeguards draw opposition not just because of concerns over costs and intrusiveness, but also because the NNWS feel that the nuclear weapon states have not lived up to their side of the NPT's grand bargain and that the burdens of the treaty are being borne disproportionately.

But it is not just the disarmament obligations that are problematic. Unadvisedly, Article IV of the NPT purports to grant all parties the 'inalienable right' to 'develop research, production and use of nuclear energy for peaceful purposes without discrimination'. But that right is not inalienable, even within the terms of the NPT itself, since it is subject to compliance with the treaty and with nuclear safeguards. Article IV also does not commit any particular state to share its particular nuclear technology with any other. But these niceties tend to be ignored by the more radical developing states that rail against verification, export controls and the alleged stinginess of technical assistance in the peaceful uses of nuclear energy. Among the states that take this tack from within the regime are Cuba, Egypt and Iran, while India and Pakistan have long berated it from without. More moderate states like Algeria, Brazil and Malaysia also echo this line, as does the non-aligned group as a whole. Iran is currently seeking to take full advantage of its 'inalienable right' in arguing that there should be no constraints on its uranium enrichment programme.

Exacerbating the situation, the NPT contains a loophole: a state can acquire all elements of the nuclear fuel cycle – from uranium mining to enrichment and reprocessing – as long as it declares them and subjects them to safeguards. But on six months' notice it may withdraw, perfectly legally, from the treaty on national security grounds and move immediately to acquire nuclear weapons. North Korea did so in 1993 (although some states dispute the legality of its withdrawal). Proposals have been made in recent years to close this loophole to prevent states from violating the treaty and withdrawing without consequence.[5]

Five-yearly NPT review conferences are typically the arena where these abiding controversies over compliance with and implementation of the NPT erupt. Some conferences are perceived to have advanced the cause of non-proliferation, such as the 2000 Review Conference, which produced the politically binding Thirteen Practical Steps to nuclear disarmament (UN 2000). The 2005 review conference ended in acrimony, without any final document. Not only the lack of progress in nuclear disarmament, but disputes about the Middle East and the Iranian and North Korean non-compliance cases played their part. The most recent review conference, held in May 2010, did achieve agreement on a final document. This was in large part due to President Barack Obama's ushering in a new US commitment to nuclear disarmament and the non-proliferation regime in general, which proved sufficient to balance the continuing stalemate over Iran and enduring issues such as the Middle East nuclear weapon-free zone. Whether this positive moment can be maintained remains to be seen. While the review conference outcome is welcome it papered over, rather than resolved, fundamental rifts among NPT parties.

Nuclear weapon-free zones

Among the most important of the additional legal instruments in the non-proliferation arena are the nuclear weapon-free zone (NWFZ) agreements that now cover a significant portion of the globe. They comprise: the 1967 Treaty of Tlatelolco, which created the world's first NWFZ, for Latin America and the Caribbean; the 1986 Treaty of Rarotonga for the South Pacific; the 1995 Treaty of Bangkok for Southeast Asia; the 1995 Treaty of Pelindaba for Africa, and the 2005 Treaty of Tashkent for Central Asia.[6]

While the zones have some variation in their provisions, mostly relating to nuclear transit, nuclear dumping and nuclear security, they largely follow a pattern. In terms of the peaceful uses of nuclear energy, all rely on IAEA safeguards to verify non-diversion to military purposes. All have putative mechanisms for dealing with allegations of non-compliance, but all are institutionally weak. While the Latin American zone has a small dedicated Secretariat and the African zone envisages establishing one, the rest rely on existing regional organizations. All of the zones have protocols open to accession by the nuclear weapon states, inviting them to provide assurances that they will respect the zone and not use nuclear weapons against zone members. While NWFZs do not substantially alter the obligations of non-nuclear weapon states party to the NPT, they do provide regional reinforcement of non-proliferation norms and compliance expectations.

Role of the International Atomic Energy Agency

The principal organizational embodiment of the nuclear non-proliferation regime is the IAEA. A UN specialized agency, it is governed by a 35-member Board of Governors (BOG), elected on a global and regional basis by a General Conference of member states. Those elected always include the 12 'quasi-permanent' members considered the most advanced in nuclear energy when the Agency was formed.[7] As an organizational instrument of global governance, the rise of the IAEA seems exemplary. Its membership has expanded from the 54 states that attended the First General Conference in 1958 to 151 members today. Its budget has increased from \$3.5 million to \$444 million (€315 million) in the same period, with an additional \$158 million (€113 million) in extra-budgetary contributions for 2010 (IAEA 2009n). The total number of support and professional staff has grown from 424 to 2,326 (IAEA 2009f; Fischer 1997: 497–498). While in the IAEA's first three years of existence it applied safeguards solely to three tons of natural uranium supplied by Canada to Japan (Fischer 1997: 82), by 2008 it had 237 safeguards agreements with 163 states, applicable to 1,131 facilities. In the same year it conducted 2,036 on-site inspections. Its Technical Cooperation (TC) programme has grown from \$514,000 in 1958 to \$194 million (€139 million) for 2010. The IAEA is regarded as one of the most efficient and well-managed UN agencies. The 2004 UN High-Level Panel on Threats, Challenges and Change said it 'stands out as an extraordinary bargain' (UN 2004: 18). In 2006 the US Office of Management and Budget gave it a virtually unprecedented rating of 100 per cent in terms of value-for-money (US Office of Management and Budget 2006).

Yet one of the legacies of the IAEA Statute that has troubling implications for the current revived interest in nuclear energy is the Agency's dual role of promoting and regulating nuclear energy. Its directors-general are obliged to be enthusiastic about the spread of nuclear power to any country that desires it, while also being harbingers of nuclear catastrophe if safety, security and safeguards are not implemented and continually strengthened. The Agency's dual mandate has also manifested itself in continuous political and budgetary battles. From the outset, developing states have broadly seen the Agency's value primarily as a provider of technical assistance, while the developed states have focused more on its verification role in preventing the proliferation of nuclear weapons (although not always as enthusiastically as might be expected). Sensing that verification could not be avoided entirely, the developing states have adopted the tactic of linking increases in the verification budget to increases in the technical cooperation programme.

Nuclear safeguards

Nuclear safeguards, which had been relatively rare and crude in the early years of the IAEA, came into their own with the advent of the NPT, which imposed an obligation on the NNWS to declare all of their nuclear materials, facilities and activities (which by definition would all be for peaceful purposes) and place them under IAEA safeguards – hence the terms 'full-scope' or 'comprehensive' safeguards. Each NPT state party is required to negotiate a bilateral Comprehensive Safeguards Agreement (CSA) with the Agency to govern the application of safeguards to that state.[8] Such safeguards seek to provide reasonable assurance of the timely detection of a 'significant quantity' of declared 'special' nuclear material (notably enriched uranium and plutonium) being diverted from peaceful uses to nuclear weapons production.[9] Verification is accomplished though nuclear accountancy, on-site inspection by a standing IAEA inspectorate, and technical means, increasingly involving continuous remote monitoring.

IAEA nuclear safeguards have become increasingly authoritative and intrusive and represent a significant voluntary surrender of sovereign national prerogatives. Despite the annual battles over budgets and the outright opposition of some member states to any improvements, considerable strengthening of safeguards has occurred over the years through a combination of accretion of new parties to the NPT, creeping tightening of safeguards requirements by the Secretariat and the BOG and periodic explosions of reform agreed by consensus in response to crises. For instance, the Indian nuclear test in 1974, although not a violation of an IAEA safeguards agreement, led to the establishment in 1975 of the 20-member Standing Advisory Group on Safeguards Implementation (SAGSI), which has subsequently recommended many technical improvements to safeguards. Notwithstanding grumblings about the cost and the perceived unfair safeguards burden on states with substantial peaceful nuclear industries like Canada, Germany and Japan, the legitimacy of the system was, until the early 1990s, increasingly accepted by Agency members and its efficacy taken for granted. The Secretariat was able to report annually to the Board that it had no indication that there had been diversion of nuclear materials from peaceful to military purposes by any state (although there have been subsequent revelations of relatively minor but still troubling violations by Egypt, Romania, Taiwan and South Korea).[10]

This complacency was shattered with the revelation after the 1990 Gulf War that Iraq had been clandestinely mounting a nuclear weapons programme in parallel with its IAEA-inspected peaceful programme. The IAEA's failure to detect Iraqi activities, located in some cases 'just over the berm' from where inspectors regularly visited, brought ridicule from those who misunderstood the limitations of its mandate and despair on the part of safeguards experts who had for years feared this outcome. The most fundamental problem was that the IAEA could only monitor and inspect materials and facilities formally declared to it by states. This provided would-be proliferators with the latitude to develop substantial undeclared nuclear capabilities undetected, either co-located with declared facilities or completely separate. A further difficulty was the Agency's reliance on nuclear accountancy as the principal tool for detecting non-compliance with safeguards and, in turn, dependence on safeguards themselves as the key tool in detecting non-compliance with the NPT. Political limitations placed on the design of nuclear safeguards in the early years had led to a presumption of compliance and a conservative safeguards culture that ultimately proved unable to detect serious violations. The Agency felt it could not use all of the powers it had acquired, including 'special inspections'; it tended to ignore unofficial information or indicators of proliferation beyond diversion, notably weaponization activities (Acton and Newman 2006) and nuclear smuggling; and it failed to take a holistic view of states' activities.

A direct consequence of the Iraq case was relatively quick agreement by the BOG on the so-called 93+2 initative, a two-part programme of strengthened safeguards. Part One comprised measures that the Board concluded the Agency already had the legal authority to undertake and which could be implemented immediately. These included requesting additional information on facilities that formerly contained safeguarded nuclear materials; increased remote monitoring of nuclear material movements; expanded use of unannounced inspections; and environmental sampling at sites to which the Agency had access. In addition, the Agency was able to expand its use of open source information, including satellite imagery (increasingly available commercially at cheap rates), as well as accepting intelligence information from member states.

Part Two of 93+2 involved negotiating a supplement to states' comprehensive nuclear safeguards agreements that would provide legal authority for further safeguards measures. It took until May 1997 for the BOG to agree on the so-called Model Additional Protocol. By this stage, the shock of Iraq was wearing off and members were resuming their previous knee-jerk reactions to reform. Nonetheless, the Protocol provides for increased transparency by extending the obligations of states to declare, report and grant on-site access to the entire range of nuclear fuel cycle activities – from mining to the disposition of nuclear waste. The Protocol also requires states to report nuclear-related equipment production, imports and exports, fuel cycle-related research and development, and plans for new facilities. This enables the IAEA to assemble a more complete picture of states' nuclear activities, as opposed to the previous one based solely on materials and facilities. Unfortunately the Additional Protocol is voluntary, making it likely that only states intent on complying will adopt one without pressure. The Agency has undertaken significant efforts to promote accession, including regional workshops, but progress has been slow (IAEA 2008b). There have been accumulating calls for the Board to make the Protocol the safeguards 'gold standard' and even make it compulsory, but there is strong opposition. In practice, increasing numbers of states are adopting a Protocol, to the point where it is starting to become the norm.

Despite the legally binding obligation of NPT state parties to have a comprehensive safeguards agreement in force, as of December 2009 there were 24 states that had not complied. These were mostly African and small island states. However, 14 of these states had at least signed a CSA and another two had had their draft agreements approved by the BOG. States cannot adopt an Additional Protocol until they have a CSA in place. As of December 2009, 93 states had an Additional Protocol in force, 34 had signed one and another eight countries' agreements had been approved by the BOG (IAEA 2010b). Currently, 64 states still have the old version of a Small Quantities Protocol (SQP), which holds safeguards in abeyance if nuclear holdings are below a certain threshold. Thirty-two states have the new version in force, five others are in the process of converting old ones to new ones and two, Jamaica and Morocco, have simply rescinded them without replacement.

In addition to strengthening safeguards, the Agency has also begun implementing 'Integrated Safeguards' in select states in order to rationalize the layers of safeguards imposed over the years, thereby increasing efficiency (and, it is hoped, effectiveness). This is partly a reward for punctilious compliance by such states with all aspects of safeguards, including the Additional Protocol, as candidates must undergo rigorous examination (and cross-examination) before qualifying. An unspoken benefit for the IAEA is that its verification resources can be devoted to more productive purposes, such as intensified state evaluation. By the end of 2009, 'savings' of approximately 800 inspector days annually, or about 10 per cent of the total, were being achieved (Muroya 2009). In any one state, savings of 30–40 per cent were possible (Muroya 2009). As of January 2010, according to IAEA sources, almost 50 states had qualified for Integrated Safeguards, including all EU member states (IAEA 2010a).

The strengthened safeguards system is a great improvement on previous arrangements, increasing considerably the costs and risks for a potential proliferator and raising confidence in the ability of the Agency to achieve timely detection. It has also, to some extent, liberated the IAEA from its past timidity, both mandated and self-imposed, and emboldened it to examine the entire range of signals of a proliferator's intentions. The agency is deliberately collecting and analysing open source information; improving its remote sensing capabilities; and accepting (with an awareness of its limitations and drawbacks) intelligence information from member states obtained through so-called National Technical Means (NTM).[11] It is also seeking to overturn some of the mechanistic aspects of inspection and other practices that in the past tended to lead to institutional blindness.

Revelations of Iran's clandestine uranium enrichment programme in 2002–2003 reinforced the conclusion that the old safeguards system, which had failed to detect the country's almost 20 years of non-compliance, was grossly inadequate. But the case also permitted the Agency to flex its newly won verification muscles by investigating evidence of weaponization and the link between Iran's military and its allegedly peaceful nuclear programme, something it previously would have felt was beyond its remit. Even though Iran has not been entirely cooperative, the extra information requirements and increased Agency powers resulting from strengthened safeguards, such as complementary access, have proved potent in providing leads for the Agency to pursue through requests for further information and follow-up inspections. Environmental sampling has also proved illuminating, as has the provision of intelligence information by member states. While strengthened safeguards have helped reveal the extent of Iranian duplicity missed by the old system, they also provide increased reassurance that in the future such non-compliance cases might be detected earlier.

Yet the strengthened safeguards system, including the Additional Protocol, still leaves the IAEA a long way from the essentially 'anytime, anywhere' verification envisaged in its Statute.[12] There is still a possibility that undeclared facilities could go undetected even with the Additional Protocol in force in a potential proliferant state. A demand for a special inspection remains an extraordinary, highly politicized option that the BOG has remained reluctant to use, even in a case like Syria, which refused to grant the Agency the access needed to clarify whether it was building a nuclear reactor before Israel bombed the site in

October 2007. This implies the need for further improvements to safeguards – an 'Additional Protocol-plus'. This need not necessarily involve more legal obligations, but better technology, streamlined knowledge management within the Agency, attention to weaponization data and further pursuit of the holistic, state-oriented approach (ICNND 2009: 91–92).

A nuclear energy revival may awaken a 'sleeper' issue that has long exercised the sharpest critics of safeguards: the fact that the current system cannot provide sufficient assurance of non-diversion of fissionable material from bulk-handling facilities, such as those involved in uranium enrichment, plutonium reprocessing, and fuel fabrication. These facilities handle such large volumes of nuclear material that significant amounts, in terms of the quantities required for an illicit nuclear device, may be unaccounted for, lodged in pipes or other equipment or subject to accounting and measurement errors. The system is also currently unable to verify overnight adaptation of enrichment and reprocessing plants from declared peaceful purposes to production of weapons-useable materials.[13] Moreover, critics like Tom Cochran claim that the IAEA's 30-year old criteria for how much nuclear material is needed to make a nuclear weapon ('significant quantity') and how much time is required to convert such materials into a bomb ('conversion time') need significant revision downwards (Cochran 2007).

If a nuclear energy revival permits increasing numbers of non-nuclear-weapon states to acquire such facilities the safeguards system risks losing credibility. Proposals for fuel banks, regional or multilateral enrichment facilities and the phasing out of the use of plutonium for civilian purposes are widely deemed to be appropriate means for dealing with the proliferation implications of these developments, but all of them imply more powerful nuclear safeguards tools beyond even today's strengthened system. In the meantime it would be useful for the IAEA to frankly tell its member states where it is unable to achieve verifiability. This will not only help relieve the agency of the perennial burden of overblown expectations but should catalyse improvements in areas where they are feasible.

Other IAEA non-proliferation activities

The discovery in 2002–2003 of a global illicit nuclear smuggling network operated by Pakistani nuclear programme director Abdul Qadeer (A.Q.) Khan gave the IAEA the impetus and licence to probe such activities for the first time, both in an attempt to unravel the A.Q. Khan case and to detect new ones. In 2004 the Agency established an 'elite investigative' group, the Nuclear Trade and Technology Analysis (TTA) Unit. The unit monitors, with the help of some states and companies, refusals of suspicious import enquiries and orders, with the aim of detecting patterns and linkages. It also maintains the IAEA's institutional memory on covert nuclear related procurement activities. In addition the agency's Safeguards Information Management directorate has two small units that have quasi-intelligence functions, one that analyses open source information and another that assesses imagery. The former head of the directorate has called for a more professional, targeted IAEA 'intelligence' capability, but many member states would be wary of such a venture (Grossman 2009).

The TTA Unit needs greater cooperation from IAEA member states and companies (it is probably receiving information on only a fraction of the cases that are actually occurring) and greater financial and personnel support if it is to realize its full potential. Although in 2006 the Agency launched an outreach programme to increase participation, only several additional states out of the 20 approached are providing information (Tarvainen 2009: 63). Charles Ferguson argues that intelligence agencies, while protecting sources and methods, could and should share more information with the IAEA, noting that 'the CIA penetrated Khan's black market but kept the IAEA in the dark about this activity for years' (Ferguson 2008). David Albright contends that the work of the TTA Unit should be integrated into the IAEA's safeguards operation to 'dramatically increase the chances of detecting and thwarting illicit nuclear trade, while improving the ability of the IAEA to detect undeclared nuclear facilities and materials' (Albright 2007).

Current state of the IAEA

Despite the IAEA's importance to international security, this apparently prized Agency has been unable to secure the necessary material and financial support that it warrants. Part of the reason is that the IAEA was unable to avoid the zero real growth budgeting imposed on all UN agencies from the mid-1980s onwards. Although this may initially have helped make the Agency 'leaner and meaner', in more recent years it began to seriously threaten its effectiveness. The financial difficulties the Agency faces are partly an outcome of success: as the number of states has increased since the end of the Cold War, notably resulting from the break-up of the Soviet Union and Yugoslavia, and as more have acquired Comprehensive Safeguards Agreements and Additional Protocols, so has the verification task increased proportionately, despite later savings through Integrated Safeguards. The Agency has also been involved in unanticipated verification exercises in South Africa, Iraq, North Korea, Libya and Iran. In addition, the Agency is cooperating with the United States and Russia in repatriating HEU from research facilities in vulnerable locations around the world as part of the US Cooperative Threat Reduction (CTR) programmes and Global Partnership Against the Spread of Weapons and Materials of Mass Destruction. The Agency's increased role in nuclear safety since Chernobyl and nuclear security since 11 September 2001 have placed further strain on its budget.

Since 1985 the IAEA has been dependent on extra-budgetary contributions, including from a non-governmental organization, the Nuclear Threat Initiative (NTI), in order to keep pace with growing demands for safeguards. Even the Agency's nuclear security programme, which should be a quintessential core function, is 90 per cent funded from extra-budgetary resources (IAEA 2008c: 29). With the support of the Bush administration, the Agency did gain a one-off increase of 10 per cent in 2003, but this was phased in over 2004–2007 (IAEA 2003d: 2).[14] In the final years of Director General ElBaradei's tenure (which ended in December 2009), there was a sense of financial crisis at the IAEA. In June 2007 he decried the Board's refusal to approve a requested increase of 4.6 per cent in the annual budget, warning that the Agency's 'safeguards function' was being 'eroded over time' (Borger 2007). In June 2008, he reportedly told the BOG that the proposed 2008 budget did not 'by any stretch of the imagination meet our basic, essential requirements', adding that 'our ability to carry out our essential functions is being chipped away' (Kerr 2007).

The financial difficulties of the IAEA have had severe implications for its infrastructure and technical capabilities, as well as the size and quality of its Secretariat, including its inspectorate and technical expertise (these will be considered further in Chapter 6). Given the Agency's current critical role in nuclear safety, security and non-proliferation and in any future nuclear energy revival, however modest, the international community's miserliness towards it seems short-sighted.

Informal non-proliferation arrangements

Some of informal non-proliferation arrangements that supplement the NPT predate the treaty, some have emerged to deal with perceived lacunae in it, while others have arisen to deal with unexpected non-proliferation threats, such as the legacy of the former Soviet weapons programmes, that the NPT is ill-equipped to deal with. Other informal arrangements have arisen to avoid controversies or outright opposition that would arise if they were proposed through the formal channels of the IAEA such as the General Conference or BOG.

To begin with, there is a web of bilateral nuclear supply agreements between states that impose tougher conditions than the normal IAEA safeguards on exported materials and equipment. For example, uranium exporters like Australia and Canada have long imposed conditions of supply that prohibit retransfers of material to third countries and seek repatriation of materials in case of breach. The United States has an elaborate system for controlling the export of nuclear and dual-use technology.

There are also two programmes of activities developed specifically to deal with the legacy of the Soviet Union's former nuclear weapons programme, the various US Cooperative Threat Reduction activities and the Global Partnership Against the Spread of Weapons and Materials of Mass Destruction. These are vital in helping secure and dispose of nuclear weapons and materials; destroying former production facilities; and retraining former scientists from Soviet WMD programmes. More recently they have been extended to other states, including Ukraine and Georgia. They make a significant contribution to nuclear safety, security and non-proliferation from the standpoint of past activities, but are of limited relevance, so far, to the nuclear energy revival.

Two informal bodies have been established by nuclear supplier states to collectively strengthen nuclear export controls. The Zangger Committee (named after its inaugural Swiss chairman), which began meeting in 1971, seeks agreement among its now 36 members on nuclear material and equipment that may be exported to another NPT state party under IAEA safeguards. It produces lists of items that 'trigger' the application of safeguards. Much more controversial is the 46-member Nuclear Suppliers Group (NSG), established in 1974 after India's nuclear weapon test. It seeks to establish, by consensus, guidelines for nuclear and nuclear-related (dual-use) exports to any state, including non-NPT parties. Its self-selected membership is mostly Western, with the significant additions of Argentina, Brazil, China and Russia. NSG guidelines are implemented by each participating state in accordance with its own national laws and practices. An example is an agreement to export nuclear and dual-use items only to states which are NPT parties and have comprehensive safeguards agreements. Nuclear technology importers have long chafed at the NSG restrictions, arguing that they breach the spirit if not the letter of their 'inalienable right' to the peaceful uses of nuclear technology under the NPT. Such opposition means that the NSG remains a political lightning rod that can never be integrated into the formal structures of the IAEA.

Controversially, the NSG agreed in 2009, after much dissension among its members, to exempt India from its existing rules, in order to facilitate finalization of the 2005 US–India Nuclear Agreement (Huntley and Sasikumar 2006). Supporters of the exemption argue that it brings India partly into the non-proliferation regime by putting all of its civilian nuclear fuel cycle under IAEA safeguards. It also subjects India to political and normative pressures to induce it to adopt other non-proliferation and disarmament obligations. However, neither the US agreement nor the NSG exemption decision committed India to taking key non-proliferation steps such as signing and ratifying the Comprehensive ban treaty (CTBT) or agreeing to a ban on the production of fissionable material for weapons purposes. Critics also contend that the deal grants legitimacy to yet another state possessing nuclear weapons, opens the door to demands from Pakistan and Israel for equal treatment, frees up India's limited domestic uranium resources for its weapons programme and undermines the raison d'être of the NPT.

Another controversial informal mechanism is the Proliferation Security Initiative (PSI), proposed by the United States in 2003 to prevent the illicit shipment by air, sea or land of WMD-related materials and technologies and related delivery systems, including those pertaining to nuclear weapons. US officials consistently declare that the PSI is 'an activity, not an organization' which 'does not have a headquarters, an annual budget, or a secretariat' (Winner 2005: 129). PSI is instead a voluntary collaborative arrangement established through a Statement of Interdiction Principles by its 11 original members (Winner 2005: 130). The PSI now claims 95 participants (US Department of State 2009b). Some states, particularly China, object to the PSI and its application to ships transiting the high seas, regarding it as a threat to the Law of the Sea. Unlike slavery or piracy, they argue, international transfers of WMD are not totally proscribed under international law; furthermore, interdiction of dual-use technologies is contentious, as such equipment has both civilian and military applications and shipments may be completely innocent.

The 2003 interdiction of the German-owned ship *BBC China* with centrifuge components destined for Libya is often attributed to PSI, but was instead part of a separate effort to disrupt the Khan network (Boese 2005). Since then several individual searches have been conducted under the PSI rubric, but due to the necessary secrecy of such operations it is difficult to gauge their success. The PSI's most notable accomplishments are reciprocal ship boarding agreements concluded among PSI states[15] and international training exercises.

In 2006, President George W. Bush sought to address the problem of sensitive nuclear technology exports and the nuclear waste issue by initiating the Global Nuclear Energy Partnership (GNEP). Under GNEP, advanced nuclear energy states would supply non-nuclear-weapon states with third-generation nuclear reactors and nuclear fuel and take back the resulting spent fuel. In return such states would agree to the highest non-proliferation standards and to not engage in enrichment or reprocessing. The advanced nuclear states would retain their monopoly on such enrichment and reprocessing technologies, reprocessing spent fuel in new facilities using allegedly 'proliferation-resistant' technologies yet to be developed. The United States convened an international GNEP forum to seek states' agreement to the original GNEP principles, as well as consulting widely on civilian nuclear energy and its fuel cycle implications. Several US allies, including Australia, Canada and South Africa, objected to surrendering their right to such fuel cycle activities as enriching uranium, as well as being required to take back nuclear waste from overseas fuel sales. The original GNEP principles were modified accordingly. As of January 2010 GNEP had 25 members. The Obama administration has eliminated funding for the domestic aspects of GNEP, while internationally it has not yet formally announced a new policy. In the meantime, GNEP gatherings are continuing. A ministerial meeting in Beijing in October 2009 agreed to review future directions, including the possibility of a name change to International Nuclear Energy Framework (ICNND 2009: 142).

A lower-key but complementary US activity to GNEP is the Next Generation Safeguards Initiative (NGSI) that has been underway in the US Department of Energy's national Nuclear Security Administration since 2008. It describes itself as a 'robust, multi-year program to develop the policies, concepts, technologies, expertise, and international safeguards infrastructure necessary to strengthen and sustain the international safeguards system as it evolves to meet new challenges over the next 25 years' (NNSA 2009). NGSI seeks to build on existing partnerships with the IAEA, the Argentine–Brazilian Agency for Accounting and Control (ABACC) and leading countries in the safeguards field and to conduct outreach to states with 'credible' plans to develop nuclear energy. NGSI replicates the GNEP model in convening international meetings of partners. NGSI promises to make a concrete contribution to strengthening nuclear safeguards that should be of great assistance to the IAEA. It is worthy of emulation by others with long safeguards experience such as Australia, Canada, Germany and Japan.

Conclusions

Since the terrorist attacks of 11 September 2001 there has been heightened concern that nuclear power plants or other facilities may make tempting targets for terrorists, while nuclear materials may be purloined for use in nuclear weapons or radiological weapons or RDDs. The nexus between nuclear security and preventing the spread of nuclear weapons has thus become clearer than ever. One difficulty in dealing with nuclear security threats through global governance measures is that this area is considered the exclusive preserve of sovereign states in a way that nuclear safety, for instance, is not. As nuclear security and radiological protection measures necessarily involve key national functions such as law enforcement and control over access to information, states are understandably reluctant to expose their security and law enforcement practices to external scrutiny, let alone anything resembling external regulation.

The international nuclear security regime, if it can even be so described, is not yet ready for any form of nuclear revival that goes much beyond the existing nuclear energy states. It is newer and much less developed than those for safety and non-proliferation (although related to and mutually reinforcing of both). As in the case of nuclear safety, many (although not all) existing nuclear energy states are well practised at ensuring security for their nuclear materials and facilities. Incidents have been rare. But a rapid nuclear energy expansion risks catastrophe unless governance, both national and international, deals with nuclear security threats competently and effectively. Fortunately the heightened awareness of the need for nuclear security that the Obama administration has engendered in respect of legacy nuclear materials and facilities may spill over into the civilian nuclear energy field.

As to non-proliferation, the link between civilian nuclear energy and nuclear weapons has been an abiding one since the dawn of the nuclear age. The international non-proliferation regime, based on the IAEA, its nuclear safeguards system and the NPT, has to date prevented the spread of nuclear weapons to scores of states, but has not prevented proliferation entirely. India, Israel and Pakistan remain outside the NPT and there have been several cases of non-compliance with the treaty. The IAEA safeguards system that verifies compliance has been considerably bolstered since the early 1990s, notably via the Additional Protocol, following revelations that Iraq had come close to acquiring nuclear weapons undetected. Several other multilateral initiatives have been taken to bolster the regime, especially in response to the discovery of the A.Q. Khan nuclear smuggling network.

The regime presently faces serious challenges, notably continuing non-compliance by Iran and North Korea, non-cooperation from Syria, and by concerns that nuclear smuggling might be continuing or occur in the future. Concessions made to nuclear-armed India by the Nuclear Suppliers Group have weakened the incentives built into the regime. Not all NPT parties have safeguards in force despite their legal obligation to do so and many are still resisting the Additional Protocol. The IAEA is under-funded and under-resourced in the safeguards area, and faces critical personnel shortages, deteriorating infrastructure and progressively outdated technology. Finally, the international community has still not resolved the central contradiction of the non-proliferation regime: that some states have accorded themselves the right to retain nuclear weapons apparently in perpetuity, while all others are under legally-binding obligation never to acquire them. Such challenges to the regime will only be made worse by a careless nuclear energy revival that fails to act on the lessons learned from the original spread of peaceful nuclear technology and permits new states (and/or non-state actors) to acquire nuclear weapons.

6 Implications of a nuclear revival for global governance

The implications for global nuclear governance of the less-than-dramatic increase in nuclear energy projected by this study are obviously not as alarming as they would be if a full-bore revival were imminent. Yet even if the nuclear energy revival is confined largely to the existing nuclear energy states and a small group of newcomers, there will be growth in the numbers of nuclear reactors and fuel cycle facilities, more nuclear transport, both domestically and internationally, and more spent fuel and nuclear waste. The implications for global governance are thus sufficiently serious to warrant attention now, especially as many aspects of the nuclear regime, as we have seen, are not optimally effective today or are under threat. Indeed, the slow pace of nuclear energy expansion gives the international community breathing space to put in place the necessary reform of global governance arrangements.

This chapter considers the potential impact of a nuclear energy revival on global nuclear governance in the three key areas of safety, security and non-proliferation. The effects may be of two broad types. First, if a nuclear revival results in a serious accident, a major terrorist incident or nuclear weapons proliferation this will have an impact on the credibility and integrity of the respective safety, security and non-proliferation regimes. As in the past, such crises may ultimately lead to reform and rejuvenation, but in the short run the regimes will appear more fragile and less authoritative. This chapter therefore begins with an assessment in each case of the impact of a revival on nuclear safety, security and non-proliferation per se. The second type of impact that a revival may have is a more direct one, affecting the capacity of the three regimes to deal with increasing numbers of nuclear energy states, increasing numbers of vendors, associated companies and national regulatory authorities, more reactors and other nuclear facilities and rising volumes of nuclear material being produced, shipped, used and stored. In global governance terms, the burden will mostly fall on the IAEA but also on the treaties, mechanisms and other arrangements that have been outlined in Chapters 4 and 5. Such effects will be political, institutional and financial. This chapter will thus consider the additional demands and strains that a nuclear energy revival is likely to place on the global governance enterprise. The overall impact on the IAEA, as the key multilateral component of the global nuclear governance regime, will be dealt with in a final section.

Nuclear safety

The impact that a nuclear energy revival will have on the global nuclear safety regime will obviously depend on how well the regime is supported politically and substantively. But it will also depend on the existing nuclear energy states continuing to pursue the highest safety standards for their current fleet of nuclear reactors – whether they are refurbished or up-rated or simply operated to the end of their planned lifetime and decommissioned. In addition it will depend on how the existing nuclear energy states go about ensuring the safety of their own 'new build' and of the reactors that they design, manufacture and export to the new-comers. As to the newcomers, safety will depend on how well they are drawn into the web of treaties, peer review processes and assistance mechanisms. A nuclear revival will also have implications for the management of spent fuel and nuclear waste, nuclear transport and nuclear liability. The impact will differ depending on the type of country engaging in nuclear activity – whether experienced old hands or newcomers. While it is impossible to quantify the impact of a nuclear revival on global nuclear safety because it is unclear precisely how large that revival is likely to be, it is possible to identify some qualitative implications.

A challenge that nuclear safety and its governance faces in a nuclear revival, one that does not confront nuclear security or non-proliferation in quite the same way, is that nuclear safety is indivisible – a nuclear accident anywhere is a nuclear accident everywhere. A major nuclear accident could sap confidence in global governance of the industry and kill the nuclear revival. As Socolow and Glaser put it, 'Safety makes all plants mutual hostages' (Socolow and Glaser 2009: 35). WANO Chairman William Cavanaugh III warned the organization's biennial meeting in Chicago in September 2007 that 'Another Chernobyl or another Three Mile Island ... would be enough to halt the nuclear renaissance, with all the imaginable negative consequences to our world's economies and to the environment' (Weil 2007) – not to mention the nuclear industry itself.

Existing nuclear reactor safety: upgrades and life extensions

Many existing nuclear energy states are upgrading, up-rating and/or granting life extensions of 20–40 years to existing nuclear plants, some of which are reaching the end of their initially planned 30–40 year life span. But as Richard Meserve notes, 'ageing plants present unique safety challenges because plants and equipment can deteriorate over time through mechanisms that may not yet be fully understood', requiring 'heightened attention over time to surveillance, preventive maintenance, and component replacement' (Meserve 2009: 100–111). Additional concerns about this aspect of the nuclear revival include:

- the maintenance of safety margins at upgraded and refurbished facilities;
- succession planning for retiring staff, technical support organizations and regulatory bodies familiar with old plant types;
- finding spare parts for old reactor types or re-manufacturing them;

- continued attention to human factors and operator training; and
- ultimate safe decommissioning of old facilities.

An illustration of the difficulties is the challenge of obsolescent analogue systems, which either need to be replaced by digital ones at the risk of system malfunction, or maintained and used by new generations of operators unfamiliar with them. In 2009 the NRC asked US nuclear power plant operators to 'voluntarily' upgrade their connections to the 16-year-old Emergency Response Data System, which reports plant conditions (such as reactor core and coolant conditions and radioactivity release rates) to the Commission (Poulson 2009). The upgrade would replace telephone dial-up modems with direct computer links. As of May 2009 only 19 operators had expressed a willingness to do so. On the spare parts issue, former Inspector-General for Electricité de France, Pierre Wiroth, has claimed that the company has a 'years-long dearth of spare parts' (Brett 2009).

There is a danger that governments, utilities and international organizations will be so fixated on new build and new entrants that they will fail to pay adequate attention to the challenges of refurbishing and maintaining an increasingly old nuclear fleet. For example, an elaborate US licence renewal programme introduced by the NRC in 1991 was 'scathingly criticized' by the commission's in-house safety auditor, the Office of the Inspector General, for lacking proper documentation and failing to independently verify operator-supplied data (Brett 2009). In August 2009 the NRC issued a safety evaluation of the Indian Point nuclear plant, located 34 miles north of New York City, as part of its licence renewal process, which showed that of 87 parts of the reactor vessel and related elements examined, all but three showed signs of ageing damage, as did 39 out of 44 steam generator components and 57 of 59 structural elements (NRC 2009b). Still, the NRC concluded that Indian Point met regulatory standards for licence renewal.

Some in the industry advocate even longer life extensions than the current trend towards 20–40 year extensions, claiming that existing reactors could safely have their lives extended by 60 years, which would make some of them set for decommissioning after 100 years of operation. NRC Chairman Dale Klein warned that margins inherent in the US reactor fleet had 'allowed us to make the transition' from 40 to 60 years of operation 'fairly easily', but that operating to 80 years or beyond would require 'a very important R&D program' (MacLaclan 2009c: 3). Among the issues that need investigating are ageing of cable insulation, ageing of concrete in high-flux radiation fields and material embrittlement. He also wondered 'where are we going to get the people?'

The US experience cannot necessarily be extrapolated to other countries, some of which may be operating reactors that do not have the necessary margins to permit them to receive life extensions but which should be closed in the immediate future.

Safety of new reactor designs

Part of the sales pitch by industry and governments for a nuclear energy revival is that new, advanced designs of nuclear reactors, Generation III and Generation III+, will be inherently safer than existing models. They promise safety features built in to their designs rather than added on as in previous generations. According to the NEA some of the most common features envisaged for new generation reactors are (NEA 2008: 232):

- explicit consideration of severe accidents as part of the design basis;
- effective elimination of some severe accident sequences by the use of inherent safety features;
- significant reduction or elimination of the potential for a large radioactive release, even if a severe accident were to occur;
- improved efficiency and effectiveness of operation and maintenance through the extensive use of digital technology; and
- reduction in system complexity and avoidance of the potential for human error.

Worryingly to the layperson, however, the NEA notes that if all of these features are successfully implemented they 'could result in the reduction of on-site and off-site protective measures, such as evacuation plans for the public' (NEA 2008: 232). It is not clear that the public will be reassured by this. Since only a handful of new generation reactors are operating and most designs have not yet even been built, it is too early to assess these safety claims based on operational experience.

A key role is played, not by the international regime, but by national regulatory authorities. One reassuring feature of reactor licensing is that all national regulators insist on reviewing reactor designs before approving construction and/ or operation, even when certification or design approval (the term used in the UK) or design review (the term used in Canada) has been obtained in the vendor's country of origin or in other buyer countries. Given the political sensitivities attending the construction of nuclear power plants it is politically impossible for governments not to do so. The difficulty for the vendor is that it may have to go through virtually an entirely new process in each country in which it seeks to build a new reactor.

Certification by the US NRC is regarded by many countries as the 'gold standard', permitting reactor vendors anywhere to market their product with confidence. A recent example is that of the Westinghouse AP1000 which originally received design certification from the NRC in 2006 (WNN 2009d). The company, which plans to build some 14 reactors in the United States (and is already building two in China), submitted revisions to the certified design to 'reduce cost and financial risk to buyers, afford extra protection against large aircraft crashes, improve instrumentation and control and finalize details such as pipe layouts'. The NRC rejected the application, saying that more work was necessary on the shield building to protect the main nuclear components during events such as severe weather.

The implications of the deployment of new reactor designs for the global governance regime are several. First, the new designs may not necessarily be compatible with existing IAEA safety standards, which will need to be reviewed and revised where necessary. States with reactor vendor companies should assist the IAEA in revising its safety standards to take account of the new generation reactors since the current standards were written only with existing light water reactors in mind. For instance the IAEA's Safety Requirements document explicitly states in its introduction that it applies primarily to water-cooled reactors (Meserve 2009: 108). Since the nuclear industry also contends that small reactors will be safe and secure, such designs should be subject to the same international scrutiny as the larger units, and have with their own safety standards.

Second, the international safety regime should also be able to offer some reassurances about the safety of new designs in addition to those provided by the manufacturers and national regulators. Already the IAEA will conduct, on request, reviews of new reactors' conceptual design safety features. In July 2009 it completed such a review of the as yet unbuilt Areva-Mitsubishi 1,100 MW Atmea 1 pressurized water reactor (Nuclear News Flashes 2009d; Atmea 2009). Clearly this type of multilateral review is important in increasing confidence in the safety of new designs, not just for the benefit of sellers and purchasers but also for the peace of mind of neighbouring states which might be affected by an accident.

Third, national safety regulations should ideally be harmonized (preferably at a high level rather than according to a lowest common denominator approach) to ensure that new reactor designs are as safe as advertised and to facilitate the regulatory process across countries (WNA 2008a). Industry favours harmonization as it is likely to save money. As the WNA explains it: 'Currently, national variations in safety regulations present an obstacle to internationally standardized nuclear reactor designs, which would foster ... economies [of scale]' (WNA 2008a: 1).

Global governance initiatives have already emerged that seek to harmonize national regulatory requirements. Collaborative studies are being conducted to consider the possibility of: harmonization of codes and standards, joint inspections of manufacturers and cooperation among regulators to converge on regulatory practices for new build (MacLachlan 2009e: 6).

One approach is being pursued through the Western European Nuclear Regulators Association (WENRA), whose members are senior regulators from all of the EU states with nuclear power sectors, plus Switzerland and Italy. In January 2007 the group published 'safety reference levels' covering existing reactors and is about to commence work on requirements for new designs (WENRA 2009). Additional groups considering the issue include: the NEA Committee on Nuclear Regulatory Activities (CNRA)'s Working Group on Regulation of New Reactors (WGRNR); the Generation IV International Forum Risk and Safety Working Group; and the WNA's Working Group on Cooperation in Reactor Design

Evaluation and Licensing (Cordel). Bilaterally, the NRC is working extensively with Chinese regulators to explain the design certification for the AP1000 'so they understand the process, not just the outcome' (MacLachlan 2009d: 3).

The most impressive harmonization efforts in regulatory terms are reportedly being carried out by the Multinational Design Evaluation Program (MDEP). Initiated in 2005, its main objective is to establish 'reference' regulatory practices and regulations that could ultimately lead to a 'multinational vendor inspection program' (NEA 2008: 232–233). Currently the participants are the national regulators of the ten countries that have three-quarters of the world's operating reactors and the most new reactors under review – Canada, China, Finland, France, Japan, Russia, South Africa, South Korea, the UK and the United States. The IAEA also participates, but secretariat services are provided by the NEA. MDEP 'interfaces' with all of the groups mentioned above.

As an example of its work, MDEP's working group on the Areva EPR has reportedly identified safety issues addressed in one country but not fully considered in others (NEA 2009a: 4). This is undoubtedly a reference to the allegedly unnecessarily complicated digital control system that the UK's Nuclear Installations Inspectorate has expressed concerns about (Stellfox 2007: 7). MDEP has also established groups on Vendor Inspection Cooperation; Digital Instrumentation and Controls; Codes and Standards; and Component Manufacturing Oversight. An innovative system of parallel vendor inspections is envisaged, in which two or more regulators conduct inspections of a nuclear component manufacturer simultaneously and compare results.

These programmes indicate that key regulators are taking the challenge of safety requirements for new reactor designs seriously. The key will be to involve not just the most competent and professional regulatory bodies, but those which are inexperienced, under-resourced or subject to extraneous political and commercial pressures to quickly approve 'new build' and ensure that regulations do not hamper their operations and commercial success. Michael Micklinghoff, chairman of the Cordel group, has noted that following discussions on this issue at the NEA and IAEA in September 2009 he sees the possibility that some countries considering new reactors – especially newcomers like Italy or the UAE or those with small regulatory bodies – could endorse and use a joint design review process relatively quickly (MacLachlan 2009e: 6).

Safety of new build in existing nuclear energy states

Although states with a large civilian nuclear sector, like Canada, the United States and Russia, have had decades of maintaining and in some cases upgrading their large reactor fleets, few have built new reactors in the past 20 years. Others, like Argentina and Brazil, have only ever had one or two reactors on which to base their experience. The danger is that in their rush to expand nuclear energy, even the most experienced states may permit industry to compromise on safety under pressure of cost-overruns and lengthening construction schedules. Former NRC Commissioner Peter Bradford told a US Senate committee in March 2009 that before the Three Mile Island accident the NRC had moved too quickly to license too many reactors in the United States. He said that one of the lessons of Three Mile Island was that 'nuclear power is least safe when complacency and public pressure to expedite are highest' (Nuclear News Flashes 2009c).

In existing nuclear energy producers the major safety issues will include:

- ensuring quality control and maintenance of high safety standards during rapid, large new build programmes where several facilities are being built simultaneously;
- avoiding shortages of trained and experience safety personnel, both at nuclear plants themselves and in the nuclear regulatory authority, including at senior management levels;
- ensuring that increasing consolidation and internationalization of the nuclear industry does not dilute or undervalue nuclear experience at the top levels in the new utility companies and vendor groups (NEA 2008: 229); and
- ensuring harmonization of safety standards in a situation where different brands of nuclear power plants are being built by multinational consortia and supplied by multiple parts suppliers (the shortage of spare parts may tempt counterfeiters to enter the market as they have done in the aircraft spare parts business).

In planning ambitious new programmes, existing states may encounter a shortage of regulatory and expert safety personnel, especially as older generations of skilled personnel from the first 'wave' of nuclear energy in the 1970s and 1980s are nearing retirement. The British nuclear regulator, the Nuclear Installations Inspectorate (NII), reportedly remains chronically understaffed, according to its umbrella organization, the Health and Safety Executive, and thus is ill-prepared for assessing new reactor designs as part of the UK's planned nuclear energy expansion (Nuclear News Flashes 2007c). The December 2008 National Regulatory Review (the Stone Report) aimed to address this situation (Stone 2008). While making up such shortfalls is largely the role of the state concerned, the international regime can help by expanding training programmes, further pursuing peer review and lessons-learned processes and ensuring that the IAEA itself is staffed with the necessary expertise in nuclear and radiological safety.

WANO is attempting to prepare its members (all of which by definition are existing nuclear energy producers) for a nuclear revival, including organizing the first ever plant managers conference. One of the challenges identified by WANO is the increasingly common recruitment of senior utility executives who have no nuclear experience. Face-to-face meetings with some of them had been 'eye-opening', revealing that they 'were not aware of the weaknesses' of their plant operations (WANO 2007). Worryingly, there have been reports that WANO has had trouble engaging with senior executives of large companies, like Electricité de France and some of the German companies, which did not see much need for interacting with the organization. Chairman William Cavanaugh III noted in 2007 that record levels of nuclear safety cooperation are being achieved among the world's nuclear operators, but that the task was never-ending and would be made more challenging by the nuclear revival (WANO 2007):

> Meeting the unprecedented demands of the nuclear renaissance will require operators not only to take on their individual responsibility to guarantee the safety of their own fleet, but also to assume a collective responsibility to work together to continually upgrade the safety of operating nuclear power stations worldwide. The public demands no less from us. We have not gathered here to pat one another on the back. The test of public confidence is like a rigorous exam on the subject of safety that all of us in the nuclear field must take every day. There will never be a time when we no longer have to take the test.

In April 2010 WANO announced that the 'forthcoming nuclear expansion' requires a certain reinvention of the organization (WNN 2010a). Its goal would be to ensure that every new reactor is given a thorough, independent review, either by WANO or by the IAEA, before it starts up. WANO will gather experts with 'pre-startup' peer review experience from its regional centres into a team to be headed by a specialist at its London headquarters. Once suitable practices and documents are ready, at least one of these core team members will participate in every pre-start review. WANO is also dramatically changing its previous policy of allowing only one member per country. Now any owner or operator of a nuclear power or fuel reprocessing plant has the right to join. Managing Director Jeff Felgate noted that the change

> recognizes the fact that multi-national companies today have interests in a variety of countries and many nations are served by multiple corporations. Owners may also become members of WANO. They control the purse strings and can therefore have an impact on safety.
>
> (WNN 2010a)

Safety of new build in new nuclear energy states

Since companies in the existing nuclear energy states are the ones designing and building new generation reactors with a view to exporting them, responsibility rests with them and their governments to ensure that their new designs can be operated as safely as possible. The new entrants that buy the new reactor types, with the exception of states like Israel, Italy and Poland, are unlikely to have the capacity or experience to adequately assess the safety of their purchases. For the vast majority of aspirant states major preparatory work is necessary in the safety realm even before they begin considering embarking on a civilian nuclear energy programme. All of the governance deficits discussed in Chapter 3 as being barriers to developing countries acquiring nuclear energy – political violence, government effectiveness, corruption and regulatory capacity – are pertinent here. Nuclear aspirant Nigeria, for example, where corruption and mismanagement are endemic (*The Economist* 2009a: 30–32), has difficulty running the oil industry in which it has been significantly engaged for decades.

As for the global nuclear safety regime itself, while it may be currently effective for longstanding nuclear energy states, which have had decades to establish and refine their safety systems and processes, it remains unprepared for an expansion involving states without previous experience of nuclear energy (or of any sophisticated industrial technology).

To begin with there are yawning gaps in the coverage of the SENES states by the key safety-related international agreements (the CNS, CENNA, CACNARE and the Joint Convention – see Annex 1). Four SENES states – Bahrain, Kenya, Namibia and Venezuela – are, disturbingly, party to none of them. If they are serious about nuclear energy they clearly have a long way to go in preparing for it. While many SENES states are party to the two nuclear accident conventions, since they have few obligations to fulfil and stand to benefit in case of an accident, a surprising number are not party to either the CNS or the Joint Convention. In total 13 SENES states have neither signed nor ratified the CNS. These absences from the international conventions means that such states are not drawn into review or peer review processes, they are not subject to peer pressure to improve their safety preparations or performance and they are not legally bound to comply with even the general safety principles promulgated by the IAEA. Certainly they are not sanctionable in the case of material non-compliance with the obligations of the international safety regime.

Clearly a priority is to bring all of these potential new entrant states into the relevant conventions as soon as possible before they move towards acquiring their first nuclear power reactor. Since achieving an internationally accepted level of safety is not a prerequisite to signing any of the safety-related conventions, and aspirant states have no existing unsafe nuclear reactors that would have to be shut down under Article 6 of the CNS, new entrants should be encouraged to join all of the conventions immediately.

Of course it is not enough for states to simply accede to the various conventions but to take action to comply with their requirements. Almost all of the states identified in SENES, the major exceptions being Israel and Italy (which previously had a nuclear energy programme), lack the requisite national regulatory laws and regulations, bodies and practices, trained and experienced personnel and appropriate safety culture. They will naturally lack plans and procedures for dealing with nuclear accidents. Under the CNS states are required to ensure that the 'necessary engineering and technical support in all safety related fields is available throughout the lifetime of a nuclear installation' (IAEA 1994: article 19 (v)). This implies that states which purchase reactors from others will need to make arrangements with vendors for the lifetime of the installation or develop their own national capability before purchase.

The IAEA, as already detailed, has an impressive array of programmes in place to advise states considering embarking on new build, including with respect to nuclear safety. It is willing, for example, to assist states by conducting feasibility studies, as it has done for the member states of the Gulf Cooperation

Council and Jordan, on whether states have the requisite capacities, including regulatory infrastructure, in place. Such documents as *Considerations to Launch a Nuclear Power Programme* (IAEA 2007a), *Milestones in the Development of a National Infrastructure for Nuclear Power* (IAEA 2007c) and *Evaluation of the Status of National Nuclear Infrastructure* (IAEA 2008f) are thorough and informative in setting out systematically the requirements for a successful nuclear energy programme. In addition, the Agency initiated in 2009 an Integrated Nuclear Infrastructure Review (INIR) process to provide a peer review of states' development towards nuclear power. In August 2009 Jordan became the first country to receive an INIR mission. The Agency has also developed a variation of its Knowledge Management to Assist Mission to conduct peer reviews of nuclear education and training systems and offer recommendations. The first such mission, to Malaysia, was completed in July 2009.

There is a danger, however, that the IAEA will be increasingly swamped by such demands and not be able to provide the necessary advice and assistance. The IAEA already says that it has received requests for advice and assistance from many of the 60 states it says are now considering nuclear power (IAEA 2009e). This points to the need for increased resources for the Agency in the area of safety (as well as security), an issue that will be considered later in this chapter.

The international regime should help the newly emerging nuclear energy states by expanding training programmes, further pursuing peer review and lessons-learned processes and ensuring that the IAEA itself is staffed with the necessary expertise in nuclear and radiological safety. The IAEA should continue to insist, as it already does, that all of the newcomers become party to all of the nuclear safety conventions outlined above as early as possible in the planning stages so that they can begin preparations early and avoid the mistakes of the longstanding nuclear energy states. As the IAEA's Deputy Director Yuri Sokolov has said 'Nuclear is a 100-year-long commitment and its sustainability, taking into account both natural and human resources together with the other aspects, should be considered' (IAEA 2009e).

Perhaps above all, the IAEA needs to have the resources to be able to authoritatively advise a state, on technical grounds, when in the Agency's judgement it is not advisable for that state to proceed to adopt nuclear energy. Sometimes this may be the only voice that is raised in opposition to a state's quixotic and perhaps ultimately expensive and potentially dangerous ambitions.

Safety of spent fuel and nuclear waste

Since nuclear fission produces radioactive products that can last from fractions of a second to billions of years, the decision to develop a nuclear power programme 'carries with it a responsibility to protect human health and the overall biota for more than thousands of years through safe and secure management of radioactive waste' (Ferguson and Reed 2009: 54). With a nuclear energy revival will come increased amounts of spent fuel and nuclear waste. How to handle 'legacy' wastes from past and existing programmes, as well as from future nuclear electricity generation, is an issue that requires urgent attention. This should come not just from the existing nuclear energy states (including those not even considering an expansion of their capacity), but needs special consideration by newcomers. The third review meeting of the parties to the Joint Convention on the Safety of Spent Fuel Management and on the Safety of Radioactive Waste Management in 2009 recommended that for aspirant states considering nuclear energy, the safety of spent fuel and radioactive waste management should 'be taken into account from the very beginning of such considerations' (IAEA 2009l: 9). This is to avoid repeating the mistakes of the original nuclear energy states which, as the World Energy Council laments, still have not yet decided how to manage the 'back end' of the nuclear fuel cycle (Marshall 2007: 17).

For a start there is a need to bring new entrants into the Joint Convention as soon as possible. Only seven SENES states are party to the Joint Convention. But just as important is for more substantive action to be taken by all states to resolve the nuclear waste impasse by:

- implementing national policies for the long-term management of spent fuel, including disposal of high level waste and/or spent fuel;
- taking decisions on siting, construction and operation of spent fuel and radioactive waste disposal facilities;
- managing 'legacy' wastes from old programmes;
- establishing better knowledge management and providing human resources; and
- setting aside financial resources for liabilities.

Initially spent fuel and nuclear waste is likely to continue to be stored at nuclear power plant sites, both in existing nuclear energy states and in new entrants. However, ultimately this material will have to be moved for reprocessing or, more likely, long-term storage. For states with smaller nuclear energy programmes the cheapest and most appealing solution for their nuclear waste would be to ship it to regional centres. One proposed way of assisting states with small nuclear programmes in dealing with their spent fuel and nuclear waste (a number likely to increase in any nuclear revival), is the idea of regional repositories.

However the report of the Third Review Meeting of the Joint Convention declared bluntly that 'no real practical progress has been achieved up to now' on this issue and suggested further cooperation between the parties (IAEA 2009s: 5). Since it is not clear that any country will be any more willing in the future to accept other countries' nuclear waste than they have been in the past (only Russia and India have given such indications), this could be a major stumbling block to even modest nuclear energy programmes in small states.

Existing nuclear energy states with large territories, those that export reactor fuel, and those like Australia, Canada and Kazakhstan that export raw uranium should, if they are serious about ensuring the worldwide expansion of nuclear energy, consider instituting 'take back' schemes for spent fuel. This would alleviate the challenge for small states of establishing their own waste repositories and have non-proliferation benefits in removing the temptation to reprocess the spent fuel in order to obtain plutonium (either for energy or weapons purposes). The Soviet Union, and subsequently Russia, operated such a scheme with the Eastern European states to which it exported nuclear power reactors. Russia is adopting the same policy of taking back spent fuel in respect of the Bushehr reactor in Iran.

Of relevance to both existing nuclear energy states and new entrants, the Third Review Conference for the Joint Convention stressed the 'utmost importance' of involving 'stakeholders and affected communities, from the beginning' in the process of developing facilities for spent fuel and waste management. This lesson has been sorely learned by the existing nuclear energy states, as exemplified in the Yucca Mountain fiasco. While it is encouraging that new entrants are being counselled to avoid such mistakes, it is of concern that so many of the likely new entrants, for instance Egypt, Iran and Morocco, do not have societies that encourage 'stakeholder' participation in any question, much less one as sensitive as nuclear waste disposition.

Some efforts are being made by aspirant states to acquire the necessary capacities. The IAEA has adopted an integrated management approach to the upgrading of national radiation protection infrastructure with the aim of achieving 'adequate' national radiation and waste safety infrastructures in participating countries and by appointing four Regional Managers for Africa, East and West Asia, Latin America, and Europe. The Agency has been faced with overwhelming demand for such services, including from many aspirant states such as Egypt, Indonesia, Iran, Kenya, Kuwait, Libya, Malaysia, Morocco, Philippines, Thailand, Tunisia and Venezuela.

Yet even states long experienced in civilian nuclear energy are struggling. The Parliamentary House of Lords Science and Technology Committee said in a report released on 3 June 2007 that the British government's proposed institutional arrangements for managing the next phase of the country's radioactive waste activity was 'incoherent and opaque' and demanded a truly independent body be established rather than an advisory group (House of Lords 2007). The nuclear waste issue alone could threaten the nuclear revival in some states, while the inability of governments and the international community collectively to solve the problem, politically and technically, will undermine confidence in the Joint Convention and in nuclear global governance generally.

Safety of nuclear transport

The nuclear revival is also likely to pose challenges to the governance of nuclear transport. It will inevitably lead to greater amounts of bulk material being transported both domestically and globally. This is likely to include uranium, LEU, fuel assemblies, spent fuel, plutonium, MOX fuel and nuclear waste. The World Nuclear Transport Institute (WNTI), a group of 42 firms that claim to be 'committed to ensuring safe nuclear transport', notes that a revival will mean an increase in the volume of nuclear material transported internationally as demand grows.

Already difficulties are being experienced due to heightened concerns about nuclear terrorism since 11 September 2001. Shipping companies and ports have faced tighter regulations which they fear will inhibit a nuclear revival. Bernard Monot, a vice-president in Areva's logistics department, states that 'The shippers complain about the port authorities, who in turn hold the shipping lines responsible and everybody accuses heavy regulations' (Stablum 2008). Some firms are leaving the business altogether due to the difficulties of shipment and trans-shipment. Moeller Maersk, the world's largest container shipping line (measured in vessel capacity), adopted a policy of not shipping radioactive materials in April 2007 (Stablum 2008), presumably because it is not worth the effort. This may permit the entry of less responsible carriers. The IAEA has set up a committee on Denial of Shipments to try to solve bottleneck in the industry. Most of these concerns, however, focus on small shipments, especially of radioactive sources which have short half-lives and need to get to customers quickly.

From a non-proliferation standpoint the idea of regional repositories or 'take back' schemes for spent fuel are eminently sensible. However both of these arrangements would increase the amount of material that needs to be safely transported long distances. The United States already faces this prospect internally as it considers alternative sites to Yucca Mountain.

Coastal and shipping nations are taking some initiatives in regard to nuclear cargo. In 2008 a group of them, with IAEA participation, held a fourth round of informal discussions in Vienna aimed at improving 'mutual understanding, confidence building and communication in relation to safe maritime transport of radioactive material' (IAEA 2009i: 24). While it is reassuring that meetings are being held, the agenda itself suggests that these areas currently need sustained attention.

As in other areas of nuclear safety, it will be critical to urge and assist new entrants in the nuclear energy business to make plans for nuclear transport as early as possible. With their focus on buying, financing, siting and building reactors they are unlikely to have even begun to consider such issues as part of their regulatory and infrastructure planning. As Jérôme Sermage, chair of the Uranium Concentrates Industry Task Force puts it, 'It is understandable that since this is both a narrow field of interest and one that is at the very beginning of the fuel cycle, that many other participants in the nuclear fuel cycle would have given little thought' to the transport issue (Sermage 2009).

Given the absence of a single treaty under the auspices of which parties can be peer reviewed or enjoined to comply with their obligations, the IAEA needs to take on the role of ensuring that aspirant nuclear energy states join the relevant international agreements and adopt and implement the Agency's Nuclear Transport Regulations. The Agency's Technical Cooperation programme needs to be boosted to provide the required advice and assistance.

Nuclear accident liability

The nuclear liability regime is clearly unprepared for a nuclear revival. The international conventions are poorly subscribed to or not yet in force, international funds are far from adequate to cover a serious accident even of the Chernobyl variety and the private insurance industry appears reluctant to insure increasingly expensive nuclear plants and the extended coverage now expected since Chernobyl. Practically none of the SENES states is party to the liability conventions. Two African states with nuclear power ambitions, Nigeria and Senegal, have recently joined, but many other SENES countries have not, including Bangladesh, Egypt, Indonesia, Jordan, Turkey, Venezuela and Vietnam.

It will be difficult to convince new entrants to the nuclear energy business to accede to all of the necessary liability conventions if the existing nuclear energy powers show such reluctance to fully support their own creations. Moreover, if the existing nuclear energy states are struggling to maintain adequate insurance for their existing fleet they will certainly face challenges in adequately insuring their fleets of new reactors. The nuclear revival will compound the insurance problem since 'There may not be sufficient market capacity to insure nuclear operators against the increased liability amounts provided for under the new or revised conventions, at least not in all countries' (Schwartz 2006: 60). While Steve Kidd of the WNA insists that for private insurance companies 'Western-designed nuclear installations are sought after businesses because of their high engineering and risk management standards' (Kidd 2009b: 13), others are not so sanguine. According to Julian Schwartz, head of legal affairs at the OECD/NEA, 'The private insurance industry has indicated it will not be able to provide coverage for new risks' (MacLachlan 2008e: 7).

There also may be political ramifications to states seeking to limit liability for nuclear accidents. The Indian government is facing opposition from within Parliament over legislation that would limit the liability of foreign companies for nuclear accidents to just $110 million, with public funds making up any shortfall in the event of a claim (*Global Security Newswire* 2010). Such a bill is one of the final requirements for implementation of the 2005 US–India nuclear cooperation agreement: if it fails it could delay US reactor sales to India.

While wealthy new entrants like the UAE will probably be able to self-insure, others like Egypt, Jordan and Vietnam will struggle to find private insurance and will find the international regime inadequate. This situation risks leaving new entrants under-insured, both in terms of national and trans-boundary coverage, and their governments (the insurers of last resort) facing financial meltdown if an accident occurs. This in turn may make governments reluctant to approve new build and act as a further damper on the nuclear energy revival. All this points to the need for overhaul of the global nuclear accident liability arrangements.

Nuclear security

Since the terrorist attacks of September 2001 there has been heightened concern that nuclear power plants or other nuclear facilities may make tempting targets for terrorist attack, may be sabotaged with help from insiders or that nuclear materials may be purloined for use in nuclear weapons or radiological weapons. But due to the pervasive secrecy in this field it is difficult to assess the current state of nuclear security worldwide and how ready states are, in security terms, for any expansion in the nuclear energy sector. As in the case of nuclear safety, the impact of a nuclear energy revival on global governance in this sector will be both indirect (arising from the impact on nuclear security per se) as well as direct (in the demands it places on global governance treaties, institutions and mechanisms).

Security of nuclear reactors and related facilities

One of the fears concerning nuclear security in the context of a nuclear energy revival is that the new nuclear facilities, built in increasing numbers in many more countries, will make tempting targets for terrorist attack, especially given their political and technological symbolism. While attacks against well-guarded and fortified nuclear power plants might seem far-fetched, a Nuclear Policy Study Group speculated as early as 1977 that, 'Terrorists might choose the nuclear industry as a target to exploit the mystique that surrounds nuclear energy and nuclear weapons' (Keeney *et al.* 1977: 301). To date only minor incidents have occurred, but threats have been made against nuclear reactors in several countries, including Australia, Canada and the United States. Such threats range from the purely symbolic to the deliberate attempt to cause a core meltdown and release of radioactivity. Notably, there is evidence that Al Qaeda considered attacking a nuclear power plant as part of the 11 September 2001 plot.

The 1977 Nuclear Policy Study Group concluded that while modern safety features reduce the likelihood of a major terrorist incident involving a civilian nuclear reactor, 'defence-in-depth' strategies must not only take into account the chance coincidence of multiple malfunctions, but the 'deliberate simultaneous sabotage of reinforcing safety measures' (Keeney *et al.* 1977: 307). Although, according to the Study Group it would require 'technically sophisticated and knowledgeable commandos' to achieve a 'high probability of causing a large radioactive release', this would not pose 'an insuperable barrier to a group with time, resources, and determination'. The report, now more than 30 years old, considered that a serious deterrent to terrorist attacks on nuclear facilities was the likelihood that the terrorists would die. Today suicide attacks are a commonplace terrorist tactic in certain parts of the world, magnifying the risk that they might be employed against sensitive facilities like nuclear power plants.[1]

Despite the rise in awareness since 11 September 2001 that hijacked commercial aircraft are capable of being used as weapons, governments and reactor vendors appear confident that that modern nuclear reactors are physically capable of withstanding deliberate aircraft crashes. A report by the Electric Power Research Institute (EPRI) conducted at the request of the Nuclear Energy Institute (NEI) in 2002 concluded that structures that house reactor fuel at US nuclear power plants would protect against a release of radiation even if struck by a large commercial jetliner. According to the report:

> state-of-the-art computer modeling techniques determined that typical nuclear plant containment structures, used fuel storage pools, fuel storage containers, and used fuel transportation containers at US nuclear power plants would withstand these impact forces despite some concrete crushing and bent steel.[2]

Others, such as Graham Allison, are not so sanguine (Allison 2004).

Following the 11 September 2001 tragedy, the NRC initiated what it termed a 'top to bottom' review of nuclear power reactor security. In February 2009 the NRC, after much deliberation, issued a final rule that requires applicants for new power reactors to assess the ability of their reactor designs to avoid or mitigate the effects of a large commercial aircraft impact. 'This is a common sense approach to address an issue raised by the tragic events of Sept. 11, 2001', said NRC Chairman Dale Klein (NRC 2009c). The NRC required, in particular, that there be strengthening of the design of the top part of a plant's outer shield building (Weil 2009a).

Since NRC decisions are influential in setting the standards for other countries, especially as most vendors wish to attract orders in the lucrative US market, such revamped policies may be adopted by others. Westinghouse, for instance, is seeking agreement for construction in China of an AP1000 design with airplane crash mitigation features that it has added since signing its contract with the Chinese (MacLachlan and Hibbs 2009). It was anticipated that China would agree to the changes because Chinese firms included by Westinghouse in the AP1000 procurement chain would later reap the benefit of a common basic design for all projects worldwide. This emphasizes the importance of and potential for international efforts to harmonize security regulations for new reactor designs through such mechanisms as the MDEP. It is not clear, though, to what extent existing nuclear reactors in all countries are capable of withstanding aircraft crashes and to what extent governments in general are taking steps to deal with the issue. Again, a lack of transparency in this field makes it difficult to assess the security of nuclear reactors worldwide.

A related security threat to nuclear power installations is a deliberate military attack by a state. This issue is a longstanding one dating back to the Israeli attack on Iraq's Osirak reactor in 1981. It has resurfaced with another Israeli attack on an alleged nuclear reactor site in Syria in September 2007, as well as intimations for several years that Israel (and maybe even the United States) was considering attacking Iran's nuclear facilities, including its nuclear power reactor at Bushehr. In the 1980s attempts were made to include a ban on such attacks in a draft Radiological Weapons Convention (RWC) that was being negotiated in the Conference on Disarmament (CD), but this ended with the demise of the RWC negotiations themselves. Given the likelihood that Israel, now a member of the CD, would not agree to a resumption of negotiations on this issue in that forum and the near certainty of failure of any talks that actually did manage to begin in other fora, this seems an unlikely candidate for extending global governance in the nuclear security area. This does not gainsay the fact that the spread of nuclear power plants throughout the world increases the chance that during a conflict a country will seek to attack such 'high value' targets. Such attacks have implications that do not attend other types of electricity generation plants, notably the risk of trans-boundary radioactive contamination in the same vein as Chernobyl. This reinforces the need for all states to become party to and implement the nuclear accident conventions, since the same measures that a state is obliged to put in place for an accident could be deployed in case of a deliberate attack.

Seizure of nuclear materials

A second type of nuclear security threat, the theft by terrorists of nuclear material for the purpose of making a nuclear weapon or RDD, is considered one of the most significant current international security threats. Successfully stealing a militarily significant amount of plutonium or HEU would certainly remove the greatest barrier faced by terrorists in achieving their goal of obtaining a nuclear weapon. Today, as Matthew Bunn notes:

> Making a bomb does not take a Manhattan project: more than 90 percent of that 1940s-era effort was devoted to making the nuclear material, not making the bomb; and that was before the basic principles of nuclear bombs were widely known, as they are today.

> (Bunn 2009: 113)

Tom Cochran of the National Resources Defense Council told a non-governmental conference held in conjunction with the April 2010 Nuclear Security Summit in Washington, DC that given the ease with which a critical mass of HEU can be detonated, the material itself should be considered a nuclear weapon – a view that went unchallenged by any of the experts present.[3] Others are more sceptical that terrorists would be readily able to purloin HEU or then fashion it into a workable, deliverable weapon (Mueller 2010: 162–223).

Theft from a standard nuclear power reactor is unlikely since the natural or low-enriched uranium used for fuel cannot be fashioned into a nuclear weapon, quite apart from the difficulties of gaining forced entry to a reactor site. Theft of spent fuel is also unlikely since it is extremely radioactive and can be handled only with special equipment and shielding, hence its description as 'self-protecting'. The heavy casks (30–100 tons in the United States) in which nuclear waste is shipped further complicate theft. MOX also poses challenges for terrorists since the plutonium would still have to be removed from it, although this is less challenging than removing it from spent fuel. Reprocessing plants, along with breeder reactors, could be targeted for the plutonium present, although due to the difficulties and dangers of working with such material there seems to be general agreement that terrorists are more likely to seek HEU, which can be more easily fashioned into a workable weapon (the 'gun assembly' type). Hence the current emphasis on securing or removing HEU from research reactors, since it is more of a 'ready made' nuclear weapon material.

Transportation of nuclear material could be seen as one of the vulnerable points for nuclear security. Transport necessarily involves removing material from fixed, large-scale facilities with highly regularized security into environments, such as transport by road, rail and sea, where there is less predictability. However, since nuclear material in transport is mobile it can also be removed from harm's way in a manner that material in a fixed location cannot be. There have in fact been no reported attacks on civilian nuclear transport to date.

Despite the barriers, dangers and drawbacks, however, terrorists might still attempt to attack or seize civilian nuclear material during transport, including, according to Keeney (Keeney 1977: 304):

- shipments of LEU from enrichment plants to fuel fabrication plants (the LEU might be seized for a radiological weapon);
- shipments of LEU or MOX from fuel fabrication plants to reactor sites; and
- shipments of plutonium from reprocessing plants to storage sites and fuel fabrication plants.

As previously mentioned to date there have been no significant incidents, either of a safety or security nature, involving the transport of civilian nuclear material. However, as an indication of its concern the NRC in May 2009 issued new US regulations to protect MOX fuel from theft or diversion, including a requirement that users of MOX with greater than 20 per cent plutonium dioxide need 'unique and separate approval from the Commission' (Weil 2009c: 2).

Security of new build in existing nuclear energy states

Most of the existing nuclear energy states appear to have good security track records, since no significant incidents are known to have occurred. Awareness and preparedness seems to have increased since 11 September 2001, but there are no public indicators of improvement as in the case of nuclear safety post-Chernobyl. Among the nuclear energy states, those with nuclear weapons have presumably extrapolated their experience in securing their nuclear weapon establishments to their civilian nuclear sectors (in some cases these sectors are closely linked), although this should not be taken for granted. Even in the nuclear weapon sector security incidents occur, as demonstrated in 2007 when the US Air Force temporarily 'lost' several nuclear weapons on a flight from North Dakota to Louisiana (Starr 2007). The US Government Accountability Office in 2007 recommended that the same security standards should be applied at commercial nuclear sites as at US nuclear weapon sites (GAO 2007), although the Nuclear Regulatory Commission (NRC) questioned this, arguing that security measures should take into account the type, form, purpose and quantity of materials (Fox 2007).

In respect of new reactor designs, the MDEP initiative should also be helpful in inculcating the concept of 'security by design', in the same way that safety and safeguards are also to be considered part of the design process. However, new reactor vendor countries like China, India and South Korea need to be drawn into this process if it is to be universally effective. Former NRC Chairman Dale Klein has called for MDEP to initiate 'multilateral agreement' on 'common threat parameters' that nuclear regulators apply worldwide for ensuring the security of nuclear power plants from 'external aggression' (Klein 2007). He admitted that the NRC itself was 'struggling with the issue of potential aircraft impacts at the design stage of new reactors' and that other countries were considering the same. The IAEA is of course aware of the security implications of a nuclear revival, noting that it 'presents opportunities and challenges in designing and incorporating concepts of nuclear security at the earliest possible stage of development and aligning them with the principles of safety and safeguards' (IAEA 2008j: 4). Since Generation III and Generation III+ reactors have not yet been built in large numbers it is difficult to assess the extent to which they will be more secure than existing reactors. Nonetheless there are generic security practices and procedures that should be applicable to all reactors and associated facilities regardless of their age or type.

Security of new build in new nuclear energy states

The acquisition of nuclear reactors by aspirant states with no security track record and non-existent security culture will represent a significant challenge both for the states themselves and for the global nuclear security regime. Increased numbers of nuclear power reactors and associated facilities in developing countries may represent 'high value' targets for secessionist movements, other rebel forces or terrorists. They may also be tempting targets in inter-state conflict, although neighbouring states may be deterred from attacking nuclear facilities given the possibility that they may also suffer the consequences of any release of radioactivity.

As in the case of nuclear safety, many new entrants will lack the necessary security capability and experience, including the requisite legislative and regulatory framework, customs and border security, and enforcement capacity. Newcomers will take years – in some estimates at least five – to establish security infrastructure, systems and practices, and much longer to establish an acceptable security culture. Corruption and poor governance generally may make this difficult to achieve.

The need for a certain level of secrecy in this field will be a continuing challenge to global governance. States are even more secretive (often for understandable reasons) about nuclear security matters than they are about nuclear safety.

International transparency is therefore constrained and IAEA involvement less welcome. As the NEA notes,

> There is an unavoidable tension between the need to communicate sufficient information to enable policymakers and the public to understand fundamental issues regarding nuclear technology, while protecting information that either contains commercially valuable proprietary information or that, if used in a malevolent manner, could pose additional risks to public health, safety and security.
>
> (NEA 2008: 309)

It recommends a 'need to know' concept with two levels of disclosure: release of 'generic' information on safety and security policies and practices to provide a measure of transparency, while limiting public release of specific information on facilities, transportation routes and other technical and operational details to avoid compromising security. However developing countries may, in principle, be much more prickly about developed states assisting them with security matters, compared with safety, as evidenced by Pakistan's long resistance to outside help in securing its nuclear establishment, both civilian and military.

Implications for the international security regime

The international nuclear security regime, if it can even be so described, is not yet ready for any form of nuclear revival that goes much beyond existing nuclear energy states. The international conventions in this field are far from universal in adherence and application. The Amendment to the Convention on the Physical Protection of Nuclear Material is not yet in force. And all of the nuclear security treaties, while legally binding in respect of their broad provisions, leave detailed implementation up to each state party. International verification of compliance and penalties for non-compliance are unknown. Even peer review is rare and likely to be based on bilateral cooperation rather than multilateral institutions.

In terms of adherence to the international nuclear safety conventions a number of aspirant states are not party to them (see Annex 2) and the extent of compliance with them is in any event largely unknown publicly due to the lack of transparency in this field. In the case of UN Security Council Resolution 1540, it is encouraging that all SENES states have submitted at least one report to the 1540 Committee. Only 20 of them have, however, submitted more than one report and only one, Algeria, has submitted an additional report in response to UN Security Council Resolution 1810 of 25 April 2008 which called for an update from each state.

None of this engenders confidence in the ability of aspirant nuclear energy states to manage the security of nuclear facilities that they may acquire, especially since other deficits in physical and institutional infrastructure and governance, including the level of corruption and mismanagement, have implications for establishing effective nuclear security regimes. The head of the US National

Nuclear Security Administration, Thomas D'Agostino, has noted that (Schneidmiller 2008): 'We're already dealing with countries that have their own views on how they protect different quantities of what kinds of materials. Normalizing those and making sure we don't open some gaps in there is very important.' Whatever one's assessment of the threat to nuclear security from a nuclear energy revival, it makes sense to at the very least to ensure full implementation of the existing treaty obligations and nuclear security principles and guidelines proffered by the IAEA. WINS could clearly play in important role in achieving this, in cooperation with the IAEA, in the same way that WANO does in the nuclear safety field.

As in the case of nuclear safety the IAEA has numerous programmes to assist states in improving nuclear security and some aspirant states are taking advantage of them. Among the SENES states that have received assistance through the Integrated Regulatory Review Service, which deals with both safety and security, are: Algeria, Belarus, Chile, Egypt, Morocco, Thailand and Tunisia (IAEA 2008j: 10). Workshops and other forms of security-related training included national workshops on nuclear law in Malawi and Nigeria and a workshop on nuclear safety, security and safeguards in Turkmenistan. Nonetheless the IAEA clearly does not have the resources or personnel to meet all requests for nuclear security advice and assistance, and as in the nuclear safety case risks being overwhelmed if the nuclear revival gathers pace.

Again, as in the case of nuclear safety, there needs to be greater cooperation among the various stakeholders involved in nuclear security. Due to the sovereign prerogatives that states claim in this field, industry seems largely content to leave matters to governments, as it does in the case of non-proliferation. However a major security incident that released radioactivity could threaten the nuclear revival in a similar fashion to a major nuclear reactor accident. In designing new generation reactors vendors need to consider security in the same way that they consider safety, while regulators need to consider how they will apply security regulations to new facilities. There is thus a need to construct a true international, universal nuclear security regime that encompasses all interested parties – international organizations, governments and industry – since all are critical players in avoiding the adverse security implications that might arise from the spread of nuclear electricity generation capacities.

Currently nuclear security is gaining a higher profile than it has in years, which should at least bring increased awareness of the issue as it pertains to civilian nuclear power. In April 2009 in a speech in Prague, President Obama announced 'a new international effort to secure all vulnerable nuclear material around the world within four years' (BBC News 2009). In April 2010 he convened a Nuclear Security Summit in Washington, DC to garner global support for such efforts and for improving nuclear security generally. While the communiqué did mention the expansion of nuclear energy and called for strengthening of the IAEA to assist states to improve nuclear security, the focus was on the security of existing nuclear materials, mostly from weapons programmes and research reactors, not the security of civilian nuclear power reactors or the materials produced by them. Paradoxically, once the legacy problems are resolved, the weakest link in nuclear security might be the civilian nuclear fuel cycle, unless this challenge is also addressed.

The Fissile Materials Working Group has proposed a 'Next-Generation Nuclear Security Initiative' that would lay out a road map for nuclear security (FMWG 2010). This should include nuclear security in respect of existing and future civilian nuclear facilities, not just legacy issues. Whether the 2010 Nuclear Security Summit leads eventually to a truly global, participatory nuclear security regime, whether WINS or the Global Initiative to Combat Nuclear Terrorism are the kernel of such an effort, whether the IAEA itself takes up the challenge, or whether it needs to be constructed on a different basis, there is an urgent need for efforts to be made now before the nuclear revival is upon us.

Clearly, though, the global security of nuclear material and installations is only as good as its weakest link and requires sustained international attention. As the IAEA dryly recommends:

> The increasingly global nature of nuclear commerce and cascading developments in fields as diverse as transport, communications and information technology make it essential that States follow international best practice in trying to limit threats directed at nuclear material and/or facilities.
>
> (IAEA 2003a: 145)

The possibility of a nuclear revival, especially in countries with weak and corrupt governance, poor regulatory systems and 'security culture' deficits, compounds the necessity of strengthening the international nuclear security regime.

Nuclear non-proliferation

The spread of peaceful nuclear energy capabilities, critics argue, goes hand-in-hand with the proliferation of latent capacities for developing nuclear explosive devices. The earliest civilian nuclear energy programmes were certainly by-products of the first nuclear weapons programmes. Yet there were concerns from the outset that the process could work in reverse. It was feared that states would seek to acquire civilian nuclear energy as a cover for a nuclear weapons programme. The current renewed enthusiasm for nuclear electricity generation is again raising fears of a wave of 'nuclear hedging' – whereby states seek the peaceful nuclear fuel cycle so they can move quickly to nuclear weapons acquisition when required. The international regime is currently being challenged in this very manner by Iran, which is engaging in precisely the type of ambiguous, hedging behaviour that an unbridled nuclear energy revival could, in a worst case scenario, unleash.

The following analysis examines to what extent a nuclear electricity generation programme, under the current safeguards regime (a significant caveat), might contribute to a nuclear weapons capability, latent or otherwise.[4] It also examines the effect that such a possibility, as part of a nuclear energy revival, might have on the nuclear non-proliferation regime. Such effects might be negative, in the sense of fraying further the bargain implicit in the NPT, or positive in encouraging further measures to reduce the chance of nuclear weapons proliferation.

The proliferation implications of civilian nuclear energy

The purported benefits of a nuclear energy programme for a nuclear weapons programme include: gaining general nuclear expertise and experience; learning how to handle radioactive materials; getting access to fissile material; and obtaining access to 'sensitive' technologies for enrichment and reprocessing.

Nuclear expertise and experience

The extent to which a nuclear energy neophyte will gain scientific expertise and experience from obtaining a nuclear power reactor depends on the existing capabilities of the country concerned and the manner in which the reactor is acquired. There is a vast difference, in terms of the expertise and experience to be gained, between a state designing and building a new reactor from scratch and buying one from a foreign supplier. If purchased on a turnkey basis, where everything is supplied by the foreign consortium, including construction, fuel and operating personnel, and the 'keys' handed over on completion, there will be little to no local nuclear learning during construction. Even if the buyer takes over the running of the plant from the outset, this will only provide experience in operating a reactor, not necessarily in designing and building another one. Some newcomer states may even contract foreign companies to run nuclear reactors on their territory indefinitely, precluding any local nuclear learning (although national regulators would presumably need to become familiar with its operation). For instance, the UAE is not only purchasing reactors from a South Korean firm on a turnkey basis, but is contracting the firm to run the reactors over their projected lifespan of 60 years (*The Economist* 2010a: 47).

Some countries like India and South Korea have learned how to build reactors by buying and eventually reverse-engineering them, but this is a long-term project without guarantee of success and will depend in part on access to commercial proprietary information. Collaborative construction projects between vendor and buyer will offer more opportunities for industrial learning by the purchasing state, but most new entrants will by definition not be in a position to contribute design or specialized construction expertise. Learning how to build a nuclear reactor could ultimately give a state the capability to build a plutonium production reactor for making fuel for nuclear weapons (but this would have to be outside of nuclear safeguards).

But even being able to replicate an imported nuclear reactor will not assist a state to readily produce a nuclear weapon. Nuclear reactors and nuclear explosives certainly both harness the energy produced by nuclear fission. Yet the technologies of the two enterprises are essentially different and require different scientific knowledge and technical expertise. These differences are substantial barriers to a state looking to advance from designing, building and operating a nuclear reactor to designing, building and detonating a nuclear device. As Mark Fitzpatrick puts it: 'Commentators with an incomplete understanding of what it takes to build nuclear weapons often assume that the acquisition of nuclear energy could be an easy stepping stone to nuclear weapons' (IISS 2009).

Acquiring and running a nuclear power reactor (or several) would certainly add to a country's general nuclear expertise and experience, especially if it already had a foundation on which to build, such as scientific expertise and a research reactor. But it is expensive, slow and not the most effective way to proceed to acquire a familiarity with basic nuclear science and technology. States seeking nuclear expertise for the first time, especially with an eventual nuclear weapons programme in mind, are most likely to begin by sending their personnel abroad for education and training in such disciplines as physics and nuclear engineering,[5] seeking assistance from other states and the IAEA,[6] establishing their own university programmes in such disciplines and by setting up nuclear research centres equipped with research reactors. As George Perkovich notes, 'There is a tendency to talk about dual-use technology, but dual-use scientists and technologists are even more important. Civil nuclear programs, with or without a nuclear power reactor, enable the training of dual-use talent' (Perkovich 2002: 193).

In some instances states seeking nuclear power reactors may already have a head start in their capability to move to nuclear weapons development in the form of research reactors, many of which use HEU. Research reactors are common among states without nuclear power reactors (IAEA 2010c). This may make the acquisition of a nuclear power reactor moot in terms of additional research and training opportunities. Successful operation of a research reactor indicates that a country already has a basis for further research into nuclear science and engineering beyond what it would acquire from one or two power reactors.

The past is instructive in this respect. India, Israel, North Korea, Pakistan and South Africa all used peaceful nuclear education, training and technical assistance, including in some cases research reactors, to enhance their potential nuclear weapons capability. And in all cases they were helped, inadvertently or deliberately, by other advanced nuclear states, something that is much less likely to happen today. India received training and technology particularly from the United States and Canada, including a research reactor used to produce the material for its 1974 nuclear test. France provided technology and equipment to Israel in the 1950s, enabling it to build a plutonium production reactor and eventually nuclear weapons. Israel did not bother with a peaceful nuclear power programme but diverted all its resources to weapons development. North Korea received assistance from the Soviet Union, including a research reactor which produced the plutonium for its nuclear test devices. South Africa received a research reactor and the HEU to fuel it from the United States, an act viewed as the genesis of its nuclear weapons programme.[7] In fact, every case of successful nuclear weapons development since the NPT entered into force in 1970 occurred with the assistance of nuclear supplier states under the guise of the pursuit of the peaceful uses of nuclear energy – but notably not specifically nuclear electricity production. The proliferation 'near-misses' of Argentina, Brazil, Iraq and Libya exhibited the same characteristics, as does the current case of Iran.

Requests for assistance were mostly justified by these states on the basis of a general interest in the peaceful uses of nuclear energy, not on the basis of their electricity generation needs. Only Argentina, Brazil India, Pakistan and South Africa went on to generate nuclear electricity for their grids, while Israel, Iraq, Libya, North Korea and, so far, Iran, have not. The acquisition of education, training and research reactors was the critical step, not the construction of a reactor for power generation. Among all of the proliferant states, only Pakistan's nuclear programme began with the acquisition of a nuclear power reactor that was purportedly for generating electricity.

Familiarity with handling radioactive material

One specific benefit of a civilian nuclear programme is learning how to handle radioactive material. The longer the material stays in a reactor, the greater the concentration of highly radioactive fission products and transuranic elements (Keeny *et al.* 1977: 246). The radioactivity of the material is several magnitudes higher than that of material produced in a dedicated plutonium production or research reactor and thus requires special handling in removing and deposition of the spent fuel in interim or long-term storage. All of the techniques involved in handling radioactive material from a dedicated plutonium production reactor can thus be learned by operating a power reactor, at least up until the reprocessing stage (Mozley 1998: 56–63). However, diverting the plutonium from a civilian nuclear reactor and removing it from the fuel rods requires additional sophisticated techniques and technologies that are not derived from operating a power reactor. Even the high-capacity French commercial reprocessing plant reportedly had difficulty cutting up fuel rods to gain access to the plutonium (Miller 2004: 49, fn. 14).

Commercial reactor spent fuel is in fact considered to be so highly radioactive as to be 'self-protecting', deterring access to the plutonium by terrorists and unsophisticated states alike. Since the uranium enrichment path to a nuclear device requires little exposure of personnel to radiation, this arguably would be the preferred option for a proliferant state concerned about such issues (although some, like Pakistan, tried both to increase their chances of success).

Access to fissile material

States seeking to acquire a nuclear power reactor for the purposes of obtaining access to fissile material for a bomb are thus likely to be frustrated, especially when such facilities are under safeguards. There are, however, some rather extreme scenarios in which this may be possible. While there has never been an instance of a state diverting uranium or plutonium from a civilian nuclear power plant for use in a nuclear device (India and North Korea diverted plutonium from research reactors), this does not mean that it is impossible (Gilinsky *et al.* 2004).

In terms of the fuel, neither the low-enriched uranium feedstock for a light water reactor (LWR) nor the natural uranium used for a heavy water reactor of the CANDU type is suitable for a nuclear weapon. Ideally, uranium needs to be enriched to 90 per cent or higher in U-235 to be considered weapons grade, compared with the 3–5 per cent used in most light water reactors. At low enrichment levels the amount of material needed for a device to reach criticality is so large that it could not realistically be detonated, particularly at enrichment levels below 20 per cent (IPFM 2007). Nuclear devices using material with somewhat lower enrichment levels have been built by advanced weapons laboratories, but the complexity and practicality of doing so drops dramatically with the enrichment level. A non-nuclear weapon state is unlikely to be able to accomplish such a difficult technical feat.   ·

Diversion of LEU fuel from a power reactor may have advantages to a proliferant state by obviating several stages in producing HEU for a bomb. Using diverted LEU from a fresh LWR fuel load in a clandestine enrichment plant can reportedly reduce the needed plant capacity by a factor of five (Gilinsky *et al.* 2004: 9). This assumes that a neophyte nuclear energy state could also secretly build a small enrichment plant and successfully evade IAEA safeguards on its reactor and fuel. While on the face of it this is implausible, the proliferation by the A.Q. Khan network of designs for basic centrifuge technology to Iran, Libya and North Korea, along with clandestine manufacture of centrifuge parts in countries like Malaysia, argues against complacency.

Plutonium contained in the spent fuel resulting from the normal operation of nuclear power reactors is also far from ideal for building a first nuclear weapon, due to the occurrence of Pu-240, an isotope of plutonium that increases proportionately the longer the fuel is left in a reactor. Pu-240 has a high rate of spontaneous fission which makes it impossible to use in a gun-assembly type weapon (of the type dropped on Hiroshima) as it will detonate prematurely. However, despite long-held beliefs to the contrary, it is theoretically possible to use it in a crude implosion device (of the type dropped on Nagasaki) that would yield at least one or two kilotons, a quite substantial explosion. The US National Academy of Sciences and US Department of Energy (DOE) reached this conclusion in the 1990s: 'Virtually any combination of plutonium isotopes … can be used to make a nuclear weapon. In short, reactor-grade plutonium is weapons-usable, whether by unsophisticated proliferators or by advanced nuclear weapons states' (Feiveson 2004: 436). The more desirable isotope of plutonium for a reliable weapon is Pu-239, which unlike Pu-240, is least abundant when fuel is irradiated for the normal three fuel cycles lasting about 60 months. However, LWR reactor fuel does not need to be kept in the core for that length of time, but could be withdrawn before it is fully 'irradiated'. According to Gilinsky *et al.*, if the operator of a newly operating LWR unloaded its entire core after approximately eight months the plutonium in the spent fuel would be weapons-grade, with a

Pu-239 content of about 90 per cent (Gilinsky *et al.* 2004: 28). About 150 kilograms of plutonium (enough for about 30 nuclear bombs) would be produced per eight-month cycle. As he and his fellow authors put it, 'The widely debated issue of the usability for weapons of plutonium from LWR fuel irradiated to its commercial limit has diverted attention from the capacity of an LWR to produce large quantities of near-weapons grade plutonium' (Gilinsky *et al.* 2004: 9).

An LWR, under safeguards, that was using larger than normal amounts of fuel would certainly come under suspicion that it was being used to produce plutonium and the IAEA is likely to detect the diversion. Moreover, the state would have to have some means of reprocessing the plutonium. However, combined with a clandestine 'quick and dirty' reprocessing plant that some experts have claimed are technically feasible, the risk of such a diversion attempt is not zero. Gilinsky *et al.* claim that under the current safeguards regime there would be little chance of detecting the diversion and processing of the plutonium into metal and its fabrication for a weapon until it was too late (Gilinsky *et al.* 2004: 22). The International Panel on Fissile Materials agrees that such a 'quick and dirty' plant could be built outside of safeguards, with minimal, rudimentary arrangements for worker radiation protection and radioactive waste management, in a year or less (IPFM 2009: 106). Ultimately a state could of course abrogate its safeguards agreement and leave the NPT on three months' notice, turning its LEU openly into a plutonium production reactor, building a reprocessing plant or using one clandestinely constructed in advance.

Allegations are frequently made that natural uranium fuelled/heavy water moderated power reactors like the CANDU are more proliferation-prone than LWRs.[8] First, CANDU-type reactors are said to produce plutonium more 'efficiently' and in larger volume per amount of fuel. Second, unlike the LWR, such a reactor does not need to be shut down to refuel (using, instead, so-called 'on-load refuelling'), thereby making it supposedly more difficult to apply safeguards to. Third, since such a reactor uses natural uranium, it does not require an enrichment facility to provide the fuel. As many countries have natural uranium deposits, this supposedly permits them to circumvent safeguards that would be imposed on imported enriched uranium as well as avoiding the expense of building their own enrichment plant. In 1977 a US study, the Ford-Mitre nuclear policy review, concluded that the CANDU was 'more suitable for reliable weapons' than conventional LWRs (Keeny *et al.* 1977).

These claims are all contested, in particular by the designers of the CANDU.[9] First, while it is true that CANDU technology 'produces the highest amount of plutonium per unit of power output of any commercial reactor' (MacKay 1998), the difference is not stark: the percentage of Pu-239 in spent fuel at discharge is 68.4 per cent, versus 57.2 per cent for a boiling water reactor and 55.7 per cent for a pressurized water reactor (Miller 2004: 43). Because the CANDU uses a much greater mass of fuel, the plutonium is 'dilute' in its spent fuel, typically 2.6 grams of fissile plutonium per initial kilogram of uranium (Whitlock 2000). Second, despite 'on-load refuelling', the IAEA has never reported any difficulty in safeguarding CANDU reactors, although they do require extra resources.

Safeguarding small numbers of fuel elements in each partial reload is in any case arguably easier than safeguarding bulk refuelling. Modern means of continuous remote monitoring helps ensure verifiability in either case. Third, the use of natural uranium can be seen as a proliferation benefit rather than a drawback, since a potential proliferant cannot use a CANDU nuclear electricity programme to justify acquiring an enrichment capability (although the new Advanced CANDU Reactor will use 'slightly enriched uranium' which renders this argument moot). Moreover, most countries do not have their own heavy water production facilities for CANDU-type reactors, so they are reliant on imports that could be cut off if proliferation concerns arose. In short, as Bratt argues, 'There is no consensus that the CANDU is a greater threat to non-proliferation than the LWR' (Bratt 2006: 46).

Plutonium from any type of reactor thus poses a certain diversion risk. But a state bent on acquiring a nuclear weapon is more likely to attempt to build a clandestine dedicated plutonium production reactor to circumvent safeguards, as Syria is suspected of attempting to do, rather than attempt diversion from a power reactor under safeguards, which runs the high risk of being discovered. As mentioned, though, the benefit of a safeguarded peaceful nuclear energy programme is that it may provide the industrial learning for a state to go on to build and operate a plutonium production reactor outside of safeguards.

Access to 'sensitive' technologies

The biggest barrier to a neophyte nuclear energy state seeking to use either uranium or plutonium from a power reactor for a nuclear weapon – besides the already formidable one of nuclear safeguards – is the difficulty of obtaining the necessary technology for enrichment and/or reprocessing.

Enrichment and reprocessing facilities are so far not widespread. The NEA identifies 13 commercial enrichment facilities and five commercial reprocessing facilities worldwide (this excludes Iran's Natanz enrichment facility, which purports to be for peaceful purposes, but is suspected of being part of a weapons programme, as well as India's research-oriented reprocessing facilities) (NEA 2008: 57; Ramana 2009). Germany, the Netherlands and the UK enrich uranium through the jointly-owned company URENCO.

A succession of states have developed enrichment and reprocessing facilities – with greater or lesser outside assistance. Until India's nuclear test in 1974, the advanced nuclear states were remarkably lax about restricting access to training and assistance in sensitive nuclear technology (GAO 1979), a situation that does not pertain today. Pakistan, Israel and North Korea all had direct outside assistance in obtaining such technology.[10] India and South Africa exploited Canadian and US assistance to develop a reprocessing capability autonomously.[11] Although there was little direct transfer of sensitive fuel cycle technology designs or equipment, both states benefited from generous technical assistance and training, and there were only rudimentary safeguards, export controls or other constraints in place (Pilat 2007). Iraq pursued old calutron technology,

Table 6.1 Commercial sensitive fuel cycle facilities, 2009

	Enrichment	*Reprocessing*
China	X	
France	X	X
Germany	X	
India		X
Iran	X	
Japan	X	
Netherlands	X	
Pakistan	X	
Russian Federation	X	X
United Kingdom	X	X
United States	X	

Source: NEA (2008: 57).

information on which had been declassified. Brazil claims to have invented its own enrichment technology, although it is widely presumed to be based on URENCO designs provided by West Germany in the mid-1970s (Spector 1988: 258). Iran benefited from enrichment design information obtained through the A.Q. Khan network, while Libya had similar assistance but to less effect.

Open acquisition of sensitive facilities

The vast majority of aspirant nuclear energy states will today not seek to obtain sensitive nuclear technology openly, at least not in the first couple of decades of commissioning their first nuclear reactor. Any state with only one or two reactors would immediately come under suspicion if it openly attempted to build an enrichment or reprocessing facility, even if it could obtain the necessary technology. It would be difficult for such a state to plausibly argue that it needed it, since it would be wildly uneconomic. (This has not stopped Iran from arguing, implausibly, for the need for an absurd 20 additional enrichment plants, even though it has no operational power reactor, and the only one being built, at Bushehr, will use imported Russian fuel.) Economies of scale suggest that any enrichment plant servicing less than about ten 1 GW reactors would be uneconomic (Feiveson *et al.* 2008: 11). It has also been estimated that 75–100 per cent of demand for enrichment services to 2030 will be satisfied by existing capacity, while demand for reprocessing services will be completely catered for by the existing over-capacity that is likely to persist into the future (ICNND 2009: 139). France, Russia and the UK, which have the greatest commercial reprocessing capacities, have had declining numbers of customers for years.

Obtaining a nuclear power reactor does not impart any particular capability to move on to developing so-called sensitive technologies, either for the front (enrichment) or back (reprocessing) ends of the nuclear fuel cycle, so the capability would have to be acquired from abroad or indigenously developed. Emerging nuclear energy states are today unlikely to openly gain access to the technology. Transfers of sensitive technology are now tightly controlled and the controls are likely to get even tighter. The G8 countries currently have in place an informal moratorium on transfers of sensitive technologies, but this is likely to be replaced in the Nuclear Suppliers Group by a criteria-based approach that would permit only the most non-proliferation-compliant states to qualify (see below for further analysis). Even then, an importing state is likely to receive the technology only in a 'black box' – meaning it can use the technology, but not obtain access to how it works.

The larger issue is not that emerging states will seek sensitive technology in the near future, but that several of the existing nuclear energy states without such capabilities, but with ambitious plans for more nuclear reactors, may do so. Their motivations may include a perceived need for energy security or to prove their technological prowess, or simply to have access to the entire nuclear fuel cycle as an 'inherent right'. Some states may persist with such technology despite the fact that domestically it may be uneconomic (depending on how many reactors they have) and that, if they wish to enter the global commercial market, they will face high barriers to entry. States with large deposits of uranium, for instance, such as Australia, Canada, Kazakhstan and South Africa, have reserved their right in principle to enrich such material to 'add value'. But there are significant cost and technology barriers to new entrants, as Australia discovered in studying the prospects (Commonwealth of Australia 2006). Argentina and Brazil are reportedly planning a joint enrichment plant. Currently some of the existing nuclear energy states are building new enrichment plants, notably the United States, and it is likely that more will be built as demand ramps up. Of the existing nuclear energy states currently without a reprocessing capability, only South Korea (controversially) is attempting to make the case for starting up in the near future.

It is this aspect of the nuclear energy revival that holds the most dangerous implications for the nuclear non-proliferation regime, since despite the application of safeguards to such facilities, they will be able to produce the material for nuclear weapons. The IAEA has difficulty applying safeguards to such facilities due to the volume of material involved, making diversion under safeguards a distinct possibility. It is therefore imperative that a solution be found that permits access to the benefits of sensitive technology without damaging the non-proliferation regime. If additional existing nuclear energy states start acquiring the full nuclear fuel cycle, it will be impossible to dissuade the newcomers from following suit.

Clandestine development of sensitive technologies

Emerging nuclear energy states with a moderate industrial capacity may be able to develop sensitive technologies relatively independently, but today they would have to do so entirely clandestinely, drawing on their existing nuclear expertise, information in the open literature, blueprints that proliferated as a result of the A.Q. Khan network, illicit imports of materials and technology, and by engaging the services of knowledgeable foreign personnel. Direct education and training in sensitive fuel cycle technologies has declined since the 1960s as a result of proliferation concerns, although it is difficult for those providing the training to draw a sharp line between what is sensitive and what is not.[12] As noted above, tightening export controls on transfers of sensitive technology make any clandestine effort much more difficult than in the past, but the movement of expert foreign personnel is less easily restricted.

On the enrichment side, the proliferation of knowledge and even blueprints for basic gas centrifuge technology to several proliferant states and unknown other recipients may benefit future proliferators. The original URENCO centrifuge design, the one first built in Pakistan by A.Q. Khan (the P1 and P2), is the logical 'starter' technology for countries that might have trouble making more sophisticated models (Miller 2004). More machines are needed than for more advanced designs, but once the technology is mastered they can be mass produced. A report by the Nonproliferation Policy Education Center in 2004 claims that 'building and operating small, covert reprocessing and enrichment facilities are now far easier than they were portrayed to be 25 years ago' (Gilinsky *et al.* 2004: 3). A key reason is the increasing availability of centrifuge technology which permits HEU to be made with 'far less energy and in far less space than was required with older enrichment methods', notably gaseous diffusion. This also makes them harder to detect. While confidence in the IAEA's ability to detect illicit HEU production at declared plants has improved dramatically since 1995 with the introduction of sampling and analysis at plants, along with wide area environmental sampling, the detection of small undeclared plants is more difficult because of their smaller 'footprint' and likely minute radioactive emissions (Miller 2004: 38–39). The question then turns on how sophisticated a state needs to be to construct a small, hidden plant.

A developing country acquiring one or two reactors is unlikely to be able to construct and operate its own enrichment plant, clandestinely or not. Even a relatively advanced country like Iran, which has been covertly seeking a nuclear weapon option for the past 20 years, is having trouble maintaining the smooth operation of relatively basic models as well as in deploying advanced ones. Centrifuge technology is inherently difficult to master. As the International Panel on Fissile Materials (IPFM) notes, studies of national centrifuge development programmes suggest it takes 10–20 years to develop the basic, first-generation technology, although this is being reduced as key technologies for producing the precision components required are increasingly available worldwide and are being integrated into computer-controlled machine tools (IPFM 2009: 105).

Laser-isotope separation (LIS), which also has low energy requirements and is even more efficient than centrifuge technology, making it faster and easier to hide, could pose a greater future proliferation risk. In 2006, General Electric and Hitachi acquired an Australian laser enrichment process, SILEX, and is planning to build a large enrichment plant based on this process in the United States. As IPFM notes, if this succeeds other states may follow (IPFM 2009: 105). Further technological improvements in enrichment, if easy and cheap to master, could be a proliferation nightmare.

As for reprocessing, the standard technology (PUREX, for plutonium/uranium extraction) is well known and relatively simple (compared to enrichment). As Marvin Miller notes, although details about how PUREX technology is implemented in specific plants is sometimes closely held for proprietary and/or national security reasons, the basic technology was declassified for the First Atoms for Peace Conference in Geneva in 1955 (Miller 2004: 44). Since then it has been described in detail in numerous reports and books and disseminated through training programmes, including those sponsored by government agencies such as the former US Atomic Energy Commission. Even so, replicating this reprocessing technology unassisted is probably beyond the capability of all of the smaller developing states currently seeking nuclear energy for the first time. However, as Miller puts it:

> The fundamental question that needs to be addressed is whether a country with a modest industrial base and a nuclear infrastructure sufficient to operate an LWR can build and operate a clandestine plant to reprocess diverted LWR fuel using the PUREX process.
>
> (Miller 2004: 45)

US expert studies since the 1950s have reportedly demonstrated the feasibility of 'quick and dirty', small, clandestine reprocessing plants specifically for separating plutonium for weapons purposes. A 1977 study at Oak Ridge National Laboratory by Floyd Cutler, one of the developers of the PUREX technology, produced a design for a minimal LWR spent fuel reprocessing plant that would operate for just several months. It would take between four and six months to build and could produce about 5 kg of plutonium, one bomb's worth, daily (Miller 2004: 48–50). The US General Accounting Office queried some of the assumptions of the study, but not the estimated construction time. In 1996, a Sandia National Laboratories team designed a minimal reprocessing plant that could be built in about six months, with an additional eight weeks needed to produce its first significant amount of plutonium (8 kg). It suggested that six skilled and experienced people would be required, readily available from nuclear weapon states or, notably, states with nuclear power plants. Although expert opinion is by no means unanimous on the feasibility of these schemes – only American studies have been considered here – and there is continuing doubt as to how sophisticated a state would need to be to succeed in implementing them, they nonetheless should give pause. Such possibilities, however remote, indicate the need for continuous review of received wisdom about the proliferation resistance of all types of nuclear technology and of the adequacy of nuclear safeguards – especially given the likelihood of additional states acquiring nuclear energy programmes.

In conclusion, a peaceful nuclear energy programme can be part of a state's trajectory towards acquiring the wherewithal for a nuclear weapons programme, but it is neither necessary nor sufficient. The main benefit to be derived from obtaining one or more power reactors, operating under nuclear safeguards, for nuclear weapons 'hedging', is the acquisition of nuclear expertise, training, material and infrastructure that would be difficult, if not impossible, to camouflage in a secret programme. Having a civilian nuclear energy programme does not remove the significant obstacles to acquiring fissile material for a nuclear device, nor does it provide the capability to weaponize and deliver a nuclear bomb.[13] A civilian nuclear energy programme may provide some opportunities for fissile material diversion, however unlikely, rendering the spread of peaceful nuclear energy not entirely risk-free from a proliferation standpoint.

What a civilian nuclear energy programme really can provide is more ethereal than commonly suspected – a plausible cover for seeking a broad range of nuclear expertise, experience and technology, including the full nuclear fuel cycle, without arousing suspicion of nuclear weapon intentions. Since a complete ban on the use of nuclear energy for peaceful purposes is totally impractical, the role of the global non-proliferation regime, notably safeguards, is two-fold: to make misuse of and diversion from the civilian fuel cycle more difficult, time-consuming and transparent; and to detect and expose at the earliest point possible the development of a clandestine weapons programme. As in the case of the fight against global terrorism, the non-proliferation regime needs to keep ahead of the ingenuity of those who would misuse technology intended for peaceful purposes.

The proliferation risk of existing nuclear energy states

The implications of a nuclear energy revival for nuclear proliferation are unlikely to be as dramatic as many fear. States with existing civilian nuclear energy programmes have already had the opportunity to derive weapons-related benefits from their programmes had they so chosen. Some, like India, Israel, Pakistan and North Korea, have long since drawn on purportedly peaceful nuclear programmes (although not from nuclear electricity programmes) to do so. Some states with nuclear power plants, like Argentina, Brazil, South Africa and Ukraine, have either ended their nuclear weapon programmes or given up actual nuclear arsenals. Others, like Belgium, Japan and Mexico, have never had nuclear weapon aspirations. Adding to their civilian nuclear energy capabilities is unlikely to entice any existing nuclear energy states into pursuing or resuming the pursuit of nuclear weapons. All of the existing nuclear energy states, with the exception of those with nuclear weapons, are currently under strengthened, full-scope safeguards as a result of having a comprehensive safeguards agreement with the IAEA. Most have an Additional Protocol.

However, several states with significant nuclear activities have not yet concluded an Additional Protocol, including Iran and North Korea. Iran, which applied its Additional Protocol on a 'provisional basis' from December 2003, suspended its cooperation with the Agency under the agreement in 2005. In terms of the nuclear revival, it is particularly alarming that two states with significant existing civilian nuclear power programmes and plans for expansion, Argentina and Brazil, have refused to conclude a Protocol, arguing that they are already well 'safeguarded' as a result of their CSAs, their bilateral safeguards arrangement and verification agency – the Argentine–Brazilian Agency for Accounting and Control (ABACC) – and their membership of the Latin American Nuclear Weapon Free Zone. However, in rejecting the new gold standard in safeguards, they are setting a poor example to nuclear energy aspirants and calling into question the non-proliferation credentials that they have relatively recently acquired after giving up their nuclear weapon plans in the 1980s. Brazil was also worryingly slow in agreeing to safeguards for its enrichment facility. It could especially strengthen its case for great power leadership and permanent membership of the UN Security Council, and remove continuing concerns about its nuclear-powered submarine programme if it were to adopt an Additional Protocol. This would be at little additional cost (although Brazilian reluctance is reportedly due to concerns that additional verification would reveal where it obtained its centrifuge technology from).

The main proliferation risk arising from existing nuclear energy states, as intimated above, will likely come not from their acquisition of additional nuclear reactors but from moves by such states to acquire the full nuclear fuel cycle (if they currently do not have it). Having 'sensitive' capacities will not necessarily imply that states will seek nuclear weapons, but it does give states the wherewithal for 'breakout' from the non-proliferation regime if their strategic circumstances change. Hence the importance of the Additional Protocol and continued advances in nuclear safeguards to enhance verifiability of non-diversion from such facilities. It would be preferable from a non-proliferation standpoint for as few states as possible to have such an option. Hence the importance, in the longer term, of efforts to provide assurances of fuel supply and to multilateralize the fuel cycle to avoid additional states acquiring enrichment and reprocessing capacities.

For several years the NSG has sought agreement on how to strengthen measures to prevent the export of particularly sensitive elements of the fuel cycle, such as enrichment and reprocessing technology. The existing NSG guidelines simply seek 'constraint' from members. At its meeting in June 2009 the NSG agreed to a criteria-based approach for the export of sensitive fuel cycle technology, but was unable to agree on a specific set of criteria. The criteria NSG members discussed included both 'objective' criteria, such as having an Additional Protocol in effect, and 'subjective' criteria, such as the effects on regional stability of introducing sensitive fuel cycle technology. Canada has objected to a US proposal that even if criteria are met, technology would only be transferred in 'black box' mode, preventing the recipient from accessing vital information about the technology and replicating it. Brazil objects to the Additional Protocol being a condition of supply, while South Africa is loathe to see any further restrictions on fellow developing countries. India is already seeking to claim that it would be exempt from new restrictions on sensitive technologies under its newly won general exemption from NSG export controls. NSG members have nonetheless attempted to engage with non-members. In 2002, they mandated the chair to continue the dialogue with countries such as Egypt, India, Indonesia,

Iran, Malaysia, Mexico, Pakistan and Israel that 'have developed nuclear programs and are potential nuclear suppliers' (CNS 2009).

The Group of 8 (G8) countries in 2004 adopted, at US urging, an informal moratorium on enrichment and reprocessing technology exports pending agreement in the NSG. This was extended each year until 2008 when it lapsed. At its July 2009 Aquila Summit in Italy, the G8 noted that the NSG had not yet reached consensus on the issue, but agreed, pending completion of the NSG's work, to implement the NSG's November 2008 'clean text' (publicly unavailable and still not agreed) on a 'national basis in the next year' (G8 2009). The G8 was unable to reach consensus on this contentious issue at its Muskoka summit in Canada in June 2010.

The proliferation risk of new nuclear energy states

If this study's projections of the size and nature of the nuclear revival are correct, there is likely to be only a handful of successful aspirant states and they will in all likelihood only acquire one or two reactors within the coming decades. These will mostly be light water reactors that use low-enriched uranium, with perhaps a few heavy water reactors of the CANDU type.

All of the SENES states will necessarily be reactor importers (with the possible exception of Israel), although they have varying degrees of existing nuclear expertise and experience, ranging from Italy and Poland with sophisticated industrial and technological backgrounds at one end of the scale to completely inexperienced developing countries like Namibia and Senegal at the other. Just over half of SENES states, however, have at least one research reactor, and a handful – Algeria, Egypt, Indonesia, Iran, Italy and Kazakhstan – have multiple research reactors, suggesting a relatively advanced nuclear research programme.

None of the states presently aspiring to nuclear energy for the first time, with the exception of Iran and Israel, is likely to have an advanced nuclear programme with a complete nuclear fuel cycle by 2030. The vast majority are unlikely to be able to enrich their own uranium or even fabricate their own fuel, with the exception of perhaps Italy and Kazakhstan, and none is likely to be reprocessing plutonium on a sophisticated scale.

Since, with the exception of Israel, all of the SENES states (along with all other non-nuclear weapon states) are party to the NPT and all have comprehensive safeguards agreements, they will be required to apply nuclear safeguards to all of their power reactors and other peaceful nuclear activities in any nuclear energy revival. There is, in addition, likely to be strong pressure on such states, if they have not already done so, to have an Additional Protocol in place, making illicit diversion more difficult than in the past. Any examination of the proliferation implications of a nuclear energy revival must take these considerations into account. While, as noted, a state with access to spent fuel could, in theory, construct a secret, illicit 'quick and dirty' reprocessing plant outside of safeguards, with minimal and rudimentary arrangements for worker radiation protection and radioactive waste management, in a year or less (IPFM 2009: 106), this would

Table 6.2 SENES operational research reactors

State	Number of reactors
Iran	5
Italy	4
Indonesia	3
Kazakhstan	3
Algeria	2
Egypt	2
Bangladesh	1
Ghana	1
Israel	1
Libya	1
Malaysia	1
Morocco	1
Nigeria	1
Poland	1
Syria	1
Thailand	1
Turkey	1
Vietnam	1
Albania	0
Bahrain	0
Belarus	0
Jordan	0
Kenya	0
Kuwait	0
Mongolia	0
Namibia	0
Oman	0
Philippines	0
Qatar	0
Saudi Arabia	0
Senegal	0
Tunisia	0
UAE	0
Venezuela	0

Source: IAEA 2009v.

certainly be beyond the technical capabilities of all but a tiny number of SENES states. Any state doing so would of course risk the possibility of exposure.

Encouragingly, most SENES states either have signed an Additional Protocol or have one in force. However, key SENES states have not even signed one, most worryingly Egypt (which might be capable of building clandestine sensitive facilities), Saudi Arabia, Syria and Venezuela. Oman and Qatar are also missing from the list. Four SENES states – Bangladesh, Ghana, Indonesia and Poland – have so far qualified for Integrated Safeguards, signifying that their past safeguards record has been impeccable.

Eight SENES states continue to have an old Small Quantities Protocol (SQP), which holds in abeyance comprehensive safeguards obligations, including declarations and inspections, while their nuclear activities remain under a certain low threshold (IAEA 1974). Controversy over SQPs arose when Saudi Arabia, a state with nuclear energy ambitions, sought one. In September 2005, the Board directed the Agency to renegotiate with SQP states to restore at least some of the IAEA's powers, based on a revised model agreement (IAEA 2006e) that obliges them to declare all of their nuclear holdings, however small, and to institute a State System of Accounting and Control. Of the SENES states only Bahrain, Kenya and Qatar have replaced their old versions with new ones. Ideally, all states seeking nuclear energy should as soon as possible swap their SQP for an Additional Protocol.

Some SENES states are already tightly bound within additional mechanisms of the non-proliferation regime. Four SENES states are members of the NSG: Belarus, Italy, Kazakhstan and Turkey. Eighteen SENES states are participants in the PSI: Albania, Bahrain, Belarus, Italy, Jordan, Kazakhstan, Kuwait, Libya, Mongolia, Morocco, Oman, Philippines, Poland, Qatar, Saudi Arabia, Tunisia, Turkey and the UAE. However, 18 SENES states are not members of nuclear weapon-free zones, which provide additional, regional non-proliferation assurances, either because zones do not exist in their region[14] or because they have not yet joined. Apart from Israel, which has nuclear weapons and is obviously not part of a nuclear weapon-free zone, they are: Albania, Bahrain, Bangladesh, Belarus, Iran, Italy, Jordan, Kuwait, Morocco, Namibia, Oman, Poland, Qatar, Saudi Arabia, Syria, Turkey and the UAE. Morocco and Namibia have signed the Treaty of Pelindaba, but are not yet parties.

In conclusion it is likely that all SENES states, with the exception of Iran, if they succeed in acquiring nuclear power reactors, will do so in the context of nuclear safeguards. While these states will acquire further general nuclear expertise and experience that may in the distant future be useful for a nuclear weapons programme, they will certainly not acquire the beginnings of such a weapons programme per se, nor will they obtain ready access to fissionable material suitable for a nuclear weapons programme, much less a 'breakout' capability. Encouragingly, some aspiring states such as the UAE are aware of the non-proliferation implications of acquiring nuclear energy and are seeking to present themselves as non-proliferation models for others to emulate.

Several states, no matter how well disposed towards implementing nuclear safeguards, will need assistance in doing so. Naturally the IAEA will be the main source of assistance. In addition, however, NGSI is working to help states that have credible plans for nuclear power to develop their safeguards infrastructure. This includes safeguards administrative authorities and frameworks, technical capacities and sustainable human resources. International training courses in the State Systems of Accounting and Control have been organized, including for states with SQPs. The United States and Australia cooperated in a workshop in August 2009 on domestic safeguards regulations for national authorities in Thailand and Vietnam. NGSI has also held several regional workshops for states with an interest in civilian nuclear power to elaborate on the IAEA document

Milestones in the Development of National Nuclear Power Infrastructure. These have been convened in Amman, Jordan, for Egypt, Jordan, Kuwait, Oman, Qatar, the UAE and Tunisia, and in Rabat, Morocco, for Algeria, Egypt, Jordan, Morocco and Tunisia (all are SENES states). In 2010, the programme is being extended to new partners, specifically Armenia and Kazakhstan, and will seek to expand cooperation with Middle East and Gulf Cooperation Council countries through both bilateral and multilateral activities. Other bilateral cooperation projects are continuing with Argentina, Brazil, China, Indonesia, EURATOM, France and Japan. Yet despite these early efforts, if and when a nuclear revival gathers pace there will be a much greater demand for assistance from neophyte states that the global governance institutions may be hard pressed to meet. The implications for the IAEA are particularly stark, not just in the safeguards area but in the safety and security realms as well.

Implications of a nuclear revival for the IAEA

While its apparently burgeoning programmes in support of states seeking nuclear energy give the impression of an organization in rude health, in fact an external management review by a consultancy company conducted in 2002 concluded that, despite its efficient management of resources, the IAEA was showing 'signs of system stress' and could not sustain its achievements or respond to increasing demands without concomitant increases in resources.[15]

A nuclear energy revival will place extra financial and resource demands on the Agency that it is unlikely to be able to meet effectively. Each additional nuclear power plant and nuclear fuel cycle facility acquired by a non-nuclear weapon state requires the application of safeguards, as does the material used in and produced by such facilities. Adoption of the Additional Protocol by increasing numbers of states will at involve higher costs at least until savings are achieved through more widespread application of Integrated Safeguards. The application of safeguards to multiple Indian nuclear facilities alone, following the 2005 US/India nuclear accord will incur significant costs, estimated in the order of €1.2 million for the first year for each new facility (IAEA 2008l). Requests for Agency assistance in safety and security generally, emergency preparedness and the disposition of nuclear waste are likely to continue to rise, even in the absence of a major revival, as states become more conscious of best practice in these areas. Demand for its Technical Cooperation programmes has been constantly increasing, even without a nuclear energy revival. Fuel bank schemes either directly operated by the Agency or multinational operations that require Agency safeguards will require additional resources. Efforts to deal with legacy issues, involving repatriation of HEU from research reactors to Russia or the United States, with IAEA participation, will bring its own costs, as will eventual IAEA involvement in verifying the disposition of surplus stocks of weapons-grade materials by Russia and the United States. In the future, the Agency may be involved in verifying North Korea's compliance with its nuclear disarmament pledges, while verification in Iran may intensify as part of a future deal.

In 2009, with the strong support of the Obama administration, steps were taken in the right budgetary direction for 2010. In September 2009, the IAEA General Conference, unusually, approved the precise amounts requested by the Secretariat: almost €315.5 million for the regular operations budget and €102,200 for the capital budget (IAEA 2009n: viii). This is an increase of €19.2 million or 6.5 per cent for 2009, well above the current inflation rate. It does not include the Nuclear Security Fund (€19.9 million), the voluntary component of the Technical Cooperation Program (€53 million) or other extra-budgetary programmes (€40.5 million) (IAEA 2009n). By comparison, the Commission of Eminent Persons in 2008 called for increases of about €50 million annually in real terms for the regular budget over several years, although it also called for a 'detailed review of the budgetary situation and additional workloads of the Agency' (IAEA 2003d).

The Agency will also need assistance in modernizing its infrastructure and technology if it is to play its part in global governance of a nuclear energy revival. Gross under-investment arising from decades of budgetary constraints has had a deleterious impact on the Agency's facilities and equipment, which now require urgent modernization. In June 2007, Director General ElBaradei noted that the organization was forced to use an unreliable 28-year-old instrument for environmental sampling and that there had been no general implementation of wide-area environmental sampling due to the projected cost (Borger 2007).

Most noticeable of the infrastructure deficits is the poor state of the Safeguards Analytical Laboratory (SAL) at Seibersdorf outside Vienna, which analyses sensitive samples from nuclear facilities and other sites. Currently, the Agency is forced to use external national laboratories for backup analysis, which, as ElBaradei told the BOG, 'puts into question the whole independence of the agency's verification system' (IAEA 2008a: 27). Using external laboratories in Western countries permits countries like Iran, for instance, to dispute the veracity of sample analysis. Most scandalously, the IAEA operation at Seibersdorf fails to meet the safety and security standards that the Agency encourages its member states to implement. Built in the 1970s, the facility requires, according to the Agency, approximately €50 million to 'prevent a potential failure in the area, which could put the credibility of IAEA safeguards at risk' (IAEA 2008a: 27). ElBaradei presented a report to the Board in October 2007 outlining the specific critical requirements for modernizing the SAL at an estimated cost of €39.2 million through 2008–2010 (IAEA 2007d).

Keeping up with the latest advances in technology is crucial to the Agency's non-proliferation mandate since it is in a sense engaged in a 'technology race' with potential proliferators that will be seeking the latest technology to advance their aims. Hence the Agency is investing in methods for detecting uranium hexaflouride gas (UF_6), which is used in centrifuges, as well as improved environmental sampling to detect minute radioactive particles. It also needs to improve its ability to verify non-diversion from bulk handling facilities such as enrichment, reprocessing and fuel fabrication plants. In addition, there is a long-term intention to replace human inspectors, where possible and appropriate, with remote monitoring technology. Furthermore, the Agency's plans to adopt a modern 'knowledge management system' cannot be fulfilled without investment in both technology and personnel. The US Next Generation Safeguards Initiative (NGSI) is seeking to assist the Agency with some of its deficits in advance of a nuclear energy revival. It has reportedly achieved 'substantial progress' towards demonstrating and institutionalizing 'Safeguards by Design', in which safeguards are incorporated into the design of new nuclear facilities at the earliest possible conceptual stage. On the technology front, a Safeguards Technology Development sub-programme is focusing on developing advanced nuclear measurement technology, unattended and remote monitoring systems, data integration and authentication applications, and field-portable detection tools to help inspectors verify the absence of undeclared nuclear materials and activities.

In April 2009, the BOG decided to establish a Major Capital Investment Fund (MCIF) for capital investment and infrastructure renewal (such as the SAL). The €12.6 million required for 2010 is to be financed through the 2010 capital budget of just €102,200, anticipated extra-budgetary contributions (€6 million) and projected savings in operational costs (€6.5 million) (IAEA 2009n: 50). The MCIF is expected to jump to more than €30 million in 2011 when major capital expenditure is expected to begin in earnest. However, neither the extra-budgetary contributions nor the operational cost savings are assured, handing the new Director General, Yukiya Amano, a major budgetary challenge in his first year in office. This outcome stands in stark contrast to the call by the Commission of Eminent Persons in 2008 for a one-time increase of €80 million for, *inter alia*, refurbishing the SAL and for adequately funding the Agency's Incident and Emergency Response Centre (IAEA 2008c: 30).

A final area where a nuclear revival will challenge the IAEA is in that of human resources. The US Government Accountability Office has described 'a looming human capital crisis caused by the large number of inspectors and safeguards management personnel expected to retire in the next 5 years' (GAO 2005). Unfortunately this is occurring just as demands on the Agency are increasing due to revived interest in nuclear energy. Like nuclear vendors, operators and regulatory agencies, the IAEA is suffering from generational change, with 20 per cent of its inspectors due to retire in the next few years (Muroya 2009) and its Secretariat generally facing bloc retirements. Due to its participation in the UN Common System, the Agency has a retirement age of 62 years for most staff and only 60 years for a quarter of them. Even in normal circumstances the Agency faces stiff competition from industry and national regulatory bodies that can offer more attractive salary and other benefits. Under current policy, for instance, the Secretariat can only offer three-year initial contracts (extendable to five or seven years, but only in limited cases for longer). This results in major losses of institutional memory and expertise. The general worldwide shortage of educated and experienced personnel in the nuclear field will take some time to alleviate. NGSI is working to 'revitalize and expand' the 'human capital base' for international safeguards in the United States by working with US National

Laboratories and universities. Despite such worthwhile national efforts there is a need for the IAEA itself to be strengthened to meet the coming demands, rather than relying on the goodwill and generosity of individual member states.

Conclusion

An increase in the use of nuclear energy for generating electricity does not axiomatically mean an increased risk of nuclear accident, nuclear terrorist incident or additional nuclear weapon states. The threat in all three areas is difficult to quantify, especially given that the nuclear revival itself is unquantifiable. The probability of the more extreme outcomes considered above is low but not non-existent and the costs of misjudging the odds are high. There is no cause for alarm given the slow pace of the nuclear revival and the existence of substantial global governance measures already in place and functioning, by and large, effectively.

Yet it is also apparent that a nuclear energy revival is likely to have a negative impact on the global governance arrangements in all three areas of safety, security and non-proliferation unless measures are taken to strengthen them. While the likely slow pace of the likely revival gives the international community breathing space in which to put additional measures in place, steps should be taken immediately not only to cope with a modest revival but as insurance against the possibility that increasing numbers of states conclude that a crash programme in nuclear energy is necessary. A strong motivation for rapid expansion would be increasing evidence of catastrophic climate change. The final chapter in this book outlines some of the measures that are recommended to strengthen global nuclear governance.

Conclusions and recommendations

This study concludes that on balance a significant expansion of nuclear energy worldwide to 2030 faces constraints that, while not insurmountable, are likely to outweigh the drivers of such an expansion. Such barriers will simply be too daunting compared to other means of generating electricity. An increase as high as a doubling of the existing reactor fleet by 2030 as envisaged in some official scenarios seems especially implausible, given that it can take a decade of planning, regulatory processes, construction and testing before a reactor can produce electricity. Even a moderately robust revival, in which growth rates equal those of the 1970s and 1980s, nuclear reactors provide a growing percentage of electricity globally, and significant numbers of new states acquire nuclear power, is at present implausible. While the numbers of nuclear reactors will probably rise from the current number, the addition of new reactors is likely to be offset by the retirement of older plants, notwithstanding upgrades and life extensions to some older facilities. Globally, while the overall amount of nuclear-generated electricity may rise, it is likely to fall as a percentage of electricity generated as other cheaper, more quickly deployed alternatives come online.

This will pertain unless governments provide greater incentives for nuclear power, including substantial subsidies and other support for first entrants, and establish high enough prices on carbon to offset the advantages of coal and to a lesser extent natural gas. Even then, nuclear will struggle due to its rising high upfront costs, history of cost- and scheduling-overruns, industrial bottlenecks and slow industrial learning. The industry will have to prove that its Generation III technology is safer, more secure and proliferation resistant and that it can achieve economies of scale through standardization, mass production and advanced management and construction techniques. The nuclear waste issue will have to be tackled decisively to assuage public opinion. In short, despite some powerful drivers and clear advantages, a revival of nuclear energy faces too many barriers compared to other means of generating electricity for it to capture a growing market share to 2030.

As to the geographic extent of a nuclear energy revival, this study projects that an expansion in nuclear energy to 2030 will be confined largely to the existing nuclear energy producers, mostly in Asia, a select number of European states and the United States, plus a few newcomers in Southeast Asia and the Middle

East. For the vast majority of states, nuclear energy will remain as elusive as ever and a worldwide revival implausible.

The implications for global nuclear governance of this less-than-dramatic increase in nuclear energy are obviously not as alarming as they would be if a full-bore revival were imminent. Nonetheless, they are sufficiently serious to warrant attention now, especially as many aspects of the nuclear regime are currently not optimally effective or are under threat. One more significant nuclear accident, one more state that develops nuclear weapons under the guise of generating electricity or one more 9/11 but with a nuclear weapon this time, is one catastrophe too many. Fortunately, the slow pace of nuclear energy expansion gives the international community breathing space to put in place the necessary reform of global governance arrangements. The following are the critical questions:

Safety – how can we commit all current and aspiring nuclear energy states to the highest nuclear safety standards?

Security – how can we ensure nuclear material and equipment is secure everywhere and not accessible by terrorists or subject to terrorist attack?

Non-proliferation – how can we prevent a nuclear revival contributing to proliferation, especially through the spread of sensitive technologies?

Nuclear safety

Nuclear safety standards have markedly improved since the wake-up call of the 1986 Chernobyl accident and its dramatic demonstration of radioactive cross-boundary effects. Old Chernobyl-style reactors have been closed, other Soviet types retrofitted for better safety, international conventions negotiated and international standards clarified and promoted. Industry itself has become more safety conscious, aware that a major nuclear accident anywhere is a major accident everywhere and could kill the prospects for a revival. Peer review has become a significant feature of the regime. Reactor designers are reportedly seeking inherently safe designs (which require no human intervention in case of a malfunction) for new generation reactors and are attempting to strengthen them against violent external events.

Still, alarming incidents continue to occur in the current nuclear fleet, even in a well-regulated industry like that of the United States. Nuclear safety is a permanent work-in-progress and complacency and regression need to be constantly combated. In many countries a lack of transparency prevents outsiders from knowing the true state of their civil nuclear installations. A rush by existing nuclear energy states to quickly add capacity runs the risk of safety being given a lower priority (as occurred in the United States in the original nuclear energy era). The danger for new entrants is that they will be unaware of and unprepared for their safety responsibilities, have no safety culture, and be too poorly governed to enforce safety regulations.

The current global governance regime for nuclear safety, despite post-Chernobyl improvements, is complex, sprawling and based on a variety of treaties and mechanisms that have arisen in different eras to meet different needs. It does, however, now seem to have all of the necessary components in place. To cope with increased use of nuclear energy it does not need wholesale reform or major additions but rather: universal adherence to existing treaties; enhancement and rationalization of existing mechanisms; and increased human and financial resources, including for regulatory purposes. While most existing nuclear energy states are party to the main safety conventions, there are yawning gaps in adherence by aspirant states that need to be filled before they are permitted to acquire nuclear power plants. In order to achieve universalization the IAEA should launch an initiative to promote the earliest possible accession by potential new nuclear energy states to all nuclear safety-related international conventions and protocols.

A glaring exception to the comprehensiveness of the nuclear safety regime is the absence of legally binding safety agreements for fuel cycle facilities (and research reactors). Given the likelihood that the number of such facilities will increase due to a nuclear energy revival (although only in current nuclear energy states in the near future), discussions should be initiated by such states on the negotiation of new legal instruments dealing with the safety of fuel cycle facilities (and research reactors) to fill this lacuna. This would ensure that the regime is strengthened in this area, with legally binding obligations, standards and best practice, and ideally peer review, before newcomers begin to contemplate acquiring their own nuclear fuel cycle facilities.

Unlike the non-proliferation regime, the nuclear safety regime relies not on multilateral monitoring and verification but peer review – which appears to work surprisingly well. The system includes review meetings for the CNS and Joint Convention and peer review services offered by the IAEA and WANO. States are under considerable pressure during these processes to demonstrate a good safety record to their peers. Compliance is high and improvements are being made. Transparency does need to be increased, for instance by all states making their national reports to the CNS and Joint Convention more transparent to the public and by broadening their scope to include the results of OSART and WANO peer reviews. It should be impressed on nuclear energy newcomers that peer reviews and IAEA review services are essential, if not mandatory. This could be done through politically-binding decisions of the parties to the various conventions and IAEA counselling of new entrants as to their responsibilities.

Although the IAEA's OSART programme seems generally sound and useful, the IAEA itself has only a modest role in it. The Agency oversees the programme but its own personnel do not participate in the on-site visits. This could be remedied easily and would provide the Agency with greater insight into nuclear safety in its member states. Given the potential increase in the number of reactors worldwide as part of nuclear revival, the IAEA should be given more resources for such an expanded OSART programme. Considering the time-consuming nature of the OSART process and future increased demand, one could envisage the IAEA establishing a dedicated cadre of experts in the various reactor types and technologies, including new generations, to permit Agency participation in all visits.

As in other areas of global nuclear governance, there is a distance between the IAEA and industry that needs closing. In addition to the two separate peer review processes there are two separate incident notification systems, run by the IAEA and WANO respectively. The lack of integration creates duplication of effort, unnecessary expense and lost opportunities in terms of the synergistic effect of their respective lessons-learned mechanisms. Consideration should be given to increasing cooperation between the peer review systems and integration of the notification systems in order to strengthen nuclear learning overall. A start could be made by having the IAEA collaborate with WANO in its industry-led peer review process. Consideration should also be given to joint IAEA/WANO processes, including site visits. WANO already on occasion takes into account whether particular power plants have recently received an OSART visit in planning its own visits. Due precautions would need to be taken in ensuring confidentiality of proprietary information during joint visits, but the IAEA has long had effective systems in place to achieve this.

While the peer review system itself appears to function well, there are indications that the operational experience feedback and lessons-learned processes are inadequate, do not involve all states, and suffer from a lack of transparency and openness. INSAG Chairman Richard Meserve has warned that the international nuclear community needs to do much more to collect, analyse and disseminate feedback from plant operating experience, lest failure to learn from past experience 'serves to derail' the 'promise of nuclear power' (MacLachlan 2007a: 10). Among initiatives that could be taken, including those proposed by INSAG itself (INSAG 2006b), are the following:

• enhanced use of CNS and Joint Convention review meetings as a vehicle for open and critical peer review and a source of learning about the best safety practices of others;
• enhanced utilization of IAEA Safety Standards to harmonize national safety regulations;
• enhanced exchange of operating experience for improving operating and regulatory practices; and
• intensified multinational cooperation in the safety review of new nuclear power plant designs.

The Multinational Design Evaluation Program, which is seeking harmonization of regulatory approaches to new reactor designs, appears to be a particularly effective example of government/industry cooperation. This work deserves continued support from all relevant stakeholders.

Nuclear regulators are currently organized in various 'clubs' depending on the type of reactor(s) they regulate. There are also differences in safety philosophy: these should not be permitted to stand in the way of cooperation but should be used to increase mutual understanding of different approaches and, potentially, produce harmonization. One means of achieving this goal would be the establishment of a true international nuclear regulators organization with universal membership to supplement or replace the current self-appointed clubs. There is also a need for defining the requirements for regulatory independence to ensure that all states know precisely what their obligations are in this area.

The nuclear liability arrangements are in particularly poor shape despite attempts at rationalizing and integrating them. They have so few parties that some protocols have not yet entered into force years after they were negotiated and the international funds they purport to set aside are alarmingly inadequate for a major nuclear accident of even the Chernobyl variety. These arrangements should be repaired and further integrated urgently.

First, all of the existing conventions need to attract more parties, especially among the aspirant states, be brought into force and have their international funds maximized. The IAEA and the OECD/NEA should mount a campaign to increase accessions to their nuclear liability instruments to enable them to enter into force and trigger the provision of the necessary international funding; the two organizations should work together to decrease fragmentation of the regimes. Second, a more serious attempt needs to be made to rationalize the conflicting requirements of the competing regimes or at the very least to ensure that all states understand the differences and can make wise choices. Despite efforts by INLEX, the EU and the IAEA, there is a need for further work on common understandings and simplification of the overlapping legal frameworks. Third, new entrant states will need special assistance, including in the area of national implementation legislation, to ensure that their legal framework for nuclear liability is sound. Fourth, creative ways of financing insurance at the national level are required to ensure that states are complying with their legal obligations under the liability conventions. The best solution from the point of view of the 'polluter pays' principle is to ensure that as many costs of the nuclear industry as possible are internalized and that operators and vendors establish pooling arrangements and mutual insurance schemes. This will work for states with several reactors, but will leave new entrants (those with one or two units) reliant on the insurers of last resort, their governments and the international liability regime.

The IAEA's role as the global 'hub' of nuclear safety has been steadily enhanced and become paramount since Chernobyl. In addition to acting as the secretariat for all of the new safety-related conventions, its key activities are: the setting and promotion of safety standards; safety advisory missions; management of a peer review system and the provision of technical assistance in nuclear safety. It manages an extraordinary number of programmes, measures and arrangements to guide, advise and assist states. The prospects of a nuclear revival, even one restricted to the existing nuclear energy states, plus a few new entrants, will place added responsibilities and burdens on the IAEA that it will not be able to cope with unless furnished with additional resources – technical, financial and human. A flood of new entrants to nuclear energy could overwhelm the Agency and jeopardize nuclear safety worldwide.

A radical reform would be the establishment of a Global Nuclear Safety Network, led by the IAEA, involving reactor vendors, operators, regulators and all other stakeholders in nuclear safety; this should be more than just a web-based network, as currently exists, and involve strengthening the IAEA's role as an information hub.

Nuclear security

Since the terrorist attacks of 11 September 2001 there has been heightened concern that nuclear power plants or other facilities may make tempting targets for saboteurs, while nuclear materials may be purloined for use in nuclear weapons or radiological dispersal devices. A nuclear revival would increase the amounts of nuclear material produced and transported and put pressure on the export control regime by increasing the amount and frequency of nuclear-related trade.

One difficulty in dealing with such threats through global governance measures is that nuclear security is considered the exclusive preserve of sovereign states in a way that nuclear safety is not. As nuclear security and radiological protection measures necessarily involve key national functions such as law enforcement and control over access to information, states are understandably reluctant to expose their security and law enforcement practices to external scrutiny, let alone anything resembling external regulation.

The international nuclear security regime, if it can even be so described, is not yet ready for any form of nuclear revival that goes much beyond the existing nuclear energy states. It is newer and much less developed than those for safety and non-proliferation (although related to and mutually reinforcing of both). As in the case of nuclear safety, many (although not all) existing nuclear energy states are well practised at ensuring security for their nuclear materials and facilities. Incidents have been rare.

New entrants will, however, in almost all cases lack the necessary security capability and experience, including legislative and regulatory framework, customs and border security, security culture, and enforcement capacity, including rapid response. Poor governance generally, as well as corruption and crime specifically will be barriers to quickly meeting these requirements. A rapid nuclear energy expansion risks catastrophe unless governance, both national and international, deals with nuclear security threats competently.

The international conventions in this field are far from universal in adherence and application. Significant numbers of SENES states are not party to them. The Amendment to the main legal pillar of the regime, the Convention on the Physical Protection of Nuclear Material, is vital, since it will oblige states to implement physical protection measures domestically, not just during international transport as the existing Convention requires. But the amendment is not yet in force. All states, but especially those seeking nuclear energy for the first time, should be strongly urged to accede to the CPPNM and sign and ratify the Amendment to help bring it into force as soon as possible. A campaign should be mounted by the IAEA and supportive states to achieve this relatively simple step. (The biggest challenge in treaty making is usually the negotiation of new instruments, not the subsequent signature, ratification and entry into force). The most recent treaty, the International Convention for the Suppression of Acts of Nuclear Terrorism, is a useful addition to international law in this field by seeking to criminalize individual acts and ensure a degree of uniformity internationally. It should also be subject to a campaign to increase the number of states parties.

While legally binding in respect of their broad provisions, all of the nuclear security treaties unfortunately leave detailed implementation up to each state party. Verification of compliance and penalties for non-compliance are absent and even the peer review processes common in the nuclear safety area are missing. This deficit in global governance needs to be rectified. There seems to be no reason why peer review processes should not be initiated in the security field, although confidentiality considerations will need to be taken into account in a way that is unnecessary in the safety field. Peer review could start with like-minded countries that already share security information and concerns and be built upon as confidence increases. The Nuclear Security Summit held in Washington, DC in April 2010 failed to agree on an Australian proposal that it support peer review for nuclear security. Russia led the opposition. However, the Work Plan adopted used the term 'sharing experience', which can be interpreted by those who support peer review as essentially an endorsement of that approach (America.gov 2010b).

The Nuclear Summit itself was a useful device in focusing the attention of busy government leaders on the details of this somewhat rarefied topic, both in respect of nuclear weapons and civilian nuclear energy. The communiqué recognized 'States rights to develop and use nuclear energy for peaceful purposes' (although notably avoiding the contested 'inalienable right' proclaimed by the NPT) and reiterated 'the responsibility of each State for the use and management of all nuclear material and facilities under its jurisdiction' (America.gov 2010a). Support was usefully garnered for the raft of nuclear security governance mechanisms and for the IAEA and its programmes. Since the communiqué and work plan are only politically binding (and then only on the 47 states that participated in the summit), it remains to be seen whether any concrete steps will be made to implement them, at least in the short term. Even though the summit endorsed President Obama's call to 'secure all vulnerable nuclear material in four years' this seems unlikely to be achieved given the practical and financial challenges. However, the Obama administration is apparently determined to keep the momentum of the summit going, notably through continuing consultations and a follow-up conference in South Korea as early as 2012 to assess progress. Ideally this meeting should address the security of the civilian nuclear power reactors and related fuel cycle facilities more specifically and adopt measures targeted at this area.

As part of its response to the terrorist attacks of 11 September 2001, the UN Security Council adopted Resolution 1540 in April 2004 to require states to establish national implementation measures to ensure that terrorists do not acquire so-called weapons of mass destruction (including nuclear and radiological weapons). The resolution is legally binding and states must report to the Council on the steps they have taken to implement it. To date compliance has been far from universal and as a transparency measure it has proved of limited value. Additional capacity-building is required for states unable to comply. This may eventuate through the establishment of a voluntary fund as recommended by the Comprehensive Review conducted by the 1540 Committee. But since the resolution focuses on a much wider problem than nuclear materials or reactors it is not well placed to deal with nuclear security issues specifically unless it is redirected to do so. A binding resolution by the Security Council, with the IAEA authorized to monitor implementation, would be one way to enforce verifiable nuclear security standards or at least physical protection standards worldwide. It remains to be seen, however, whether developing countries would countenance further intrusion into their sovereign prerogatives by a Security Council that they accuse of 'legislating' for them. In any event Russia is likely to veto such a measure.

The IAEA, as in the nuclear safety area, already provides a huge range of services to member states to advise, guide and assist them. The Agency's Three-Year Plans, inaugurated in the wake of 11 September 2001, have been useful in focusing the various activities and in funding them through a special Nuclear Security Fund. However, funding is still insufficient and too many conditions are attached by donors. The IAEA's Illicit Trafficking Database is another useful service but suffers from insufficient participation by states. States should provide increased funding to the regular IAEA budget that deals with nuclear security and to the Nuclear Security Fund. Given the global importance of nuclear security, donor states should drop the restrictions they have imposed on their contributions to the fund to permit the IAEA to use the resources where most needed.

Because of the secrecy that surrounds nuclear security matters (much of it for understandable reasons) international transparency is less welcome than in the nuclear safety area. The Nuclear Energy Agency recommends a sensible 'need to know' concept with two levels of disclosure: release of 'generic' information on policies and practices to provide a measure of transparency, while limiting public release of specific information on facilities, transportation routes and other technical and operational details to avoid compromising security (NEA 2008: 309). This could form the basis for rethinking the balance between the conflicting values of confidentiality and transparency that may improve global governance instruments in this field, making peer review, for instance, more feasible.

As in the case of nuclear safety, there needs to be greater cooperation among the various stakeholders involved in nuclear security. Industry seems largely content to leave matters to governments, as it does in the case of non-proliferation issues. However a major security incident at a nuclear power plant could threaten the nuclear revival in a similar fashion to a major nuclear reactor accident. In designing new generation reactors, vendors need to consider security in the same way that they consider safety, while regulators need to consider how they will apply security regulations to new facilities. International cooperation seems axiomatic.

There is a need for a truly international, universal nuclear security regime that encompasses all interested parties – international organizations, governments, regulators and industry. The US/Russia Global Initiative to Combat Global Terrorism and WINS are both excellent initiatives that deserve support from all stakeholders worldwide. Whether they are the beginnings of such a global regime or whether it needs to be constructed afresh are questions that require urgent attention.

Nuclear non-proliferation

The regime presently faces serious challenges, notably continuing non-compliance by Iran and North Korea, non-cooperation from Syria, and by concerns that nuclear smuggling might be continuing or re-emerge in the future. Concessions made to nuclear-armed India by the Nuclear Suppliers Group have weakened the incentives built into the regime. Not all NPT parties have comprehensive safeguards in force despite their legal obligation to do so and many are still resisting the Additional Protocol. The IAEA is under-funded and under-resourced in the safeguards area, and faces critical personnel shortages, deteriorating infrastructure and progressively outdated technology. The discontent of the non-nuclear weapon states with the perceived inequities of the regime risked disrupting yet another NPT Review Conference, in 2010. This was narrowly avoided only because the Obama administration, in contrast to its predecessor under George W. Bush, was well disposed towards the treaty, attentive to the concerns of other states and determined to make a success of the meeting. For instance, the United States negotiated directly with Egypt on its demand for movement towards a nuclear weapon-free zone in the Middle East.

The current challenges to the NPT regime and its review process would be exacerbated by a careless nuclear energy revival that fails to act on the lessons learned from the original spread of peaceful nuclear technology. Renewed enthusiasm for nuclear electricity generation is raising fears of 'nuclear hedging' – whereby states seek the peaceful nuclear fuel cycle to facilitate eventual acquisition of nuclear weapons as a hedge against a threatening neighbour with nuclear weapon ambitions or against an existing nuclear weapon state. The most troubling current case is the Middle East, where several states fear Iran's nuclear intentions. While the illicit diversion of nuclear material from civilian reactors or related installations – or the construction of secret enrichment and reprocessing facilities or production reactors – cannot be entirely discounted, such eventualities are unlikely. All non-nuclear weapon states are required to operate all of their nuclear facilities under strengthened comprehensive nuclear safeguards and newcomers to nuclear energy will be under strong pressure to adopt an Additional Protocol. The vast majority of states will seek civilian nuclear energy for legitimate purposes.

Nonetheless several governance responses are required to ensure the continuing health of the non-proliferation regime. Most importantly, the Additional Protocol needs to become the 'gold standard' for nuclear safeguards, including as a condition of all supply of nuclear material and technology; any state seeking nuclear power plants should be expected to immediately acquire an Additional Protocol. As well as seeking universal acquisition of an Additional Protocol, further continuing improvements should be made to the measures it provides for. There is no need for a new legally binding agreement, especially since it would be politically impossible to get such a document through the IAEA Board of Governors. Even bundling further measures together as an 'Additional Protocol-plus' is too politically provocative for most developing states to abide. However further measures can be quietly taken to provide the IAEA with advanced technology, modern facilities, the highest possible levels of expertise, and clarified mandates in respect of its verification powers and compliance determinations. All aspiring nuclear energy states need to be drawn fully into the regime as soon as possible. Additional assistance should be provided to states to improve their national implementation of nuclear safeguards, since their collaboration with IAEA headquarters and its inspectors is vital to the credibility of nuclear safeguards. In this connection, the US New Generation Safeguards Initiative could be multilateralized. This could be done gradually by other safeguards champions like Canada establishing their own versions of the Initiative.

The powers of the IAEA could be enhanced in other ways. The Director General should request special inspections in serious cases of suspected safeguards violations and non-cooperation. The Board of Governors should confirm the authority of the Agency to monitor weaponization research and development activities, especially now that a precedent has been set in the case of Iran. More states should provide information to the IAEA's Trade and Technology Analysis Unit to strengthen its efforts to expose nuclear smuggling. Such efforts should be integrated into the safeguards work of the Agency so that the necessary connections are made between illicit export and import activities and states' safeguards compliance. The IAEA Secretariat should work more closely with industry, research institutes and non-governmental organizations to take advantage of their capacities and perspectives on non-proliferation. For its part the nuclear industry, especially reactor vendors, needs to take nuclear non-proliferation more seriously and ramp up involvement with the IAEA, the NEA and WINS.

Related to the efficacy of safeguards is the question of what to do about non-compliance when it is discovered. It is essential that the Board of Governors and, if necessary, the IAEA membership as a whole, clarify the meaning of safeguards 'non-compliance' with a view to declaring zero tolerance of any breach, regardless of intent. There is an argument that both South Korea and Egypt should have been found in non-compliance for their breaches, discovered in 2004 and 2005, respectively, as should Syria for failing to cooperate fully with the IAEA in its investigation of the alleged nuclear reactor construction site destroyed by Israel. Even when the Security Council sanctioned Iran in

November 2003, it did not use the word 'non-compliance', although Libya was deemed 'non-compliant' in 2004.[1]

The BOG should instruct the Safeguards Department of the Secretariat that in future it should treat 'non-compliance' as a technical term, to be decided on the facts, and presented automatically to the Board in those terms for consideration as to the action to be taken. The Board may or may not decide to confirm such a finding, but at the very least it would permit the Board to declare factually that the Director General has recommended a finding of non-compliance based on a technical judgement by the Secretariat. This may help remove at least some of the politicization that has characterized Board behaviour in recent years and head off accusations or imputations of bias on the part of the Director General and/or the Secretariat (made by Iran, Israel and the United States over the Iran case and by the United States in the Iraq case). The Board should also formally confirm that non-cooperation by a state with the IAEA represents non-compliance: the Board should make it clear that it is the state's responsibility to prove its compliance to the Agency rather than the other way around, as has traditionally been the case.

More broadly, an attempt should be made to remove the current ambiguity about whether a violation of a safeguards agreement is a violation of the NPT. There has hitherto been an almost surreal supposition that the IAEA was not concerned with attempts by states to acquire nuclear weapons per se, but only with attempts to divert fissionable material to such a purpose. Since there is no other NPT verification body but the IAEA, it has never been clear who else was charged with considering and investigating evidence of weaponization activities. This ambiguity should be removed by BOG fiat – over the likely objections of Iran and others.

More broadly still, there have been proposals for dealing with withdrawal from the NPT by a state that is in non-compliance with its safeguards obligations and by extension the NPT. The first idea is for the UN Security Council to declare that such withdrawal would be a threat to international peace and security, requiring the necessary severe response. A second is a declaration by the NPT parties that a state withdrawing from the NPT is not entitled to use nuclear materials, equipment and technology it obtained while a party to the treaty and must return these forthwith. (The ICNND has proposed a protocol to CSAs extending safeguards in perpetuity, as in the case of the IAEA–Albania Safeguards Agreement (ICNND 2009: 89).) A third proposal is for states to make it a condition of supply that in the event of withdrawal from the NPT safeguards should continue with respect to nuclear material and equipment provided, as well as on any material produced by using it. All of these ideas have merit, although a non-compliant state withdrawing from the NPT will already have crossed such a normative and legal barrier that it is unlikely to be swayed by such legal niceties, perhaps with the exception of the Security Council actually taking enforcement action under Chapter VII of the UN Charter to restore international peace and security.

The main proliferation threat from a nuclear revival comes not from the spread of nuclear power reactors, but the possibility that increasing numbers of states, lured by the dream of energy self-sufficiency and security, will seek a complete nuclear fuel cycle – from uranium mining to the enrichment of uranium and the reprocessing of spent fuel. This would give them a way of acquiring weapons-grade material that could be diverted secretly to build a bomb or be used openly for such purposes after withdrawal from the NPT on three months' notice. The spread of sensitive enrichment and reprocessing technology to new states – whether aspirant or existing nuclear energy states or those without any interest in nuclear power generation – is in fact the single greatest threat to the non-proliferation regime. Because such facilities are dual-use and have peaceful applications permitted by Article IV of the NPT, states that meet their safeguards and non-proliferation commitments have the right to acquire them for peaceful purposes. Preventing their further spread is challenging. The difficulties in stopping Iran from enriching uranium despite Security Council resolutions ordering it to do so illustrate the problem. Several approaches are possible, with varying prospects.

One approach is technology export constraints, as currently practised by NSG members in accordance with its collective decisions or by individual suppliers through additional unilateral decisions. The extreme version of this is complete technology denial. As noted above, the NSG, encouraged by the G8, is working towards strict criteria for new enrichment or reprocessing states that should limit their emergence. Even without new agreed restrictions, supplier states are not likely to transfer the technology to new states in the near future. The most telling example of this is US refusal to allow Canada access to enrichment technology unless it is 'black-boxed'. Areva itself has purchased 'black box' enrichment technology from URENCO for its future enrichment plants, including one it plans to build in the United States (Acton 2009: 53). However, technology denial is rightly perceived as politically unsustainable in the longer term due to the demands of those non-nuclear weapon states which perceive yet another case of discrimination by the nuclear 'haves' against the nuclear 'have-nots'.

A second approach is the development of proliferation-resistant technologies that permit states to have the benefits of nuclear fuel recycling without the proliferation risks. This is not a task for global governance, although bodies like the IAEA and NEA can encourage research and development of such technologies and international cooperation to that end. Attempts have been made since the dawn of the nuclear age to design technical fixes for the proliferation problem, beginning with the idea of 'denaturing' plutonium in the 1940s. Proliferation resistance involves establishing barriers, through technological means, to the misuse of civil nuclear energy programmes to produce fissile material for nuclear weapons. In the current context of a nuclear revival, companies designing Generation III and Generation III+ reactors are claiming that they are more proliferation-resistant, although the details are still unclear. But they are still mostly LWRs that use enriched uranium (as will the Advanced CANDU) and produce plutonium, so until at least 2030 the traditional non-proliferation concerns with such technology will persist (not least because current LWRs are having their lifetimes extended beyond 2030). The traditional technological solution here is the 'once through' cycle with no reprocessing and ultimate deep geologic disposal of spent fuel.

Although the use of MOX fuel helps reduce stockpiles of spent fuel and plutonium, it still relies on plutonium being reprocessed, stored and transported and thus is a proliferation risk. Ideally the use of this 'technological solution' should be phased out altogether and aspirant states should not be encouraged to engage in it. In the meantime, in countries that already have nuclear weapons fresh MOX should be protected like their nuclear weapon material. Such protection would be more difficult to ensure if MOX was routinely used in nations that do not have reprocessing plants or MOX fabrication facilities under IAEA safeguards (Garwin and Charpak 2001: 318).

Similarly, currently proposed fast 'burner' reactors, despite advantages in using uranium resources efficiently and reducing nuclear waste, do not offer proliferation resistance since they rely on weapons-grade plutonium for fuel and/or actually produce more of it. Thorium reactors depend on reprocessed U-233, which can itself be used for nuclear weapons, in addition to requiring enriched uranium or plutonium for their initial operating cycles. As Feiveson *et al.* put it:

> It would be unwise to rush into a nuclear renaissance with the technologies that are the most developed now but which would not necessarily be the most suited to a large-scale expansion of nuclear power if other reactor technologies promise significant advantages.
>
> (Feiveson *et al.* 2008)

Many of the proposed technological solutions currently being mooted for additional proliferation resistance for Generation IV reactors and for the broader fuel cycle, notably as part of the Gen IV Forum and the IAEA's INPRO, are still in the early developmental stage.[2] Such technologies include fast neutron reactors with an integrated core and no breeder 'blanket' which ensures that plutonium will not be of weapons grade (although any plutonium is potentially weapons-useable as discussed). New reprocessing technologies promise that pure plutonium will not be separated out, but remain mixed with 'self-protecting' highly radioactive actinides. Controversially, South Korea is developing 'pyroprocessing', which it claims is proliferation-resistant, but which the US disputes. The DUPIC (direct use of spent PWR fuel in CANDU) method of recycling PWR spent fuel in CANDU reactors, without separating plutonium, shows promise, but would be limited to states with both types of reactor (currently only South Korea, India and China) (ICNND 2009: 128).

Even a cursory examination of these technologies indicates that, as the ICNND starkly concluded: 'There is no magic bullet to eliminate all proliferation risk' and 'No presently known nuclear fuel cycle is completely proliferation proof: proliferation resistance is a comparative term' (ICNND 2009: 126). In the DUPIC case for instance, a state purchasing CANDUs to recycle PWR spent fuel could always switch to traditional fuel (Acton 2009: 5). Fast 'burners' can inevitably lead to interest in fast 'breeder' reactors which are obviously more proliferation-prone (Acton 2009: 52). Political and institutional barriers will always be necessary. Nonetheless, it is important that research and development continue, especially through multilateral mechanisms, as part of a longer-term effort to find technological solutions to complement the governance solutions that this study is focused on. As James Acton points out, there is nothing wrong with pursuing technological solutions, but 'a failure to appreciate fully the political dimension of nonproliferation risks makes the concept of proliferation resistance at best irrelevant and at worst counterproductive' (Acton 2009: 49).

A third approach to restricting access to sensitive technologies is known as 'multilateralization' or internationalization of the nuclear fuel cycle. The grab bag of proposals currently being considered includes some that involve little multilateralization beyond providing assurances of reactor fuel supply so that states are not tempted to obtain their own enrichment capacities. The ultimate goal of true multilateralization would see all enrichment and reprocessing facilities for peaceful purposes being under international control and ownership, as well as under IAEA safeguards.

Although such ideas date back to the dawn of the nuclear age, and the original idea that the IAEA should have a physical fuel bank, they received a fillip in 2004 when then IAEA Director General ElBaradei convened an Expert Group on Multilateral Approaches to the Nuclear Fuel Cycle. A great number of proposals have been put forward since, some restricting themselves specifically to providing guaranteed nuclear fuel, while others veer towards multilateralizing parts of the fuel cycle (see Annex 4 for a list) (Yudin 2009).

Several of the current proposals aim to discourage states from seeking enrichment technology simply by providing them with assurances of fuel supply in all circumstances except violation of safeguards and/or the NPT. Such assurances are meant to convince states that they will not be deprived of nuclear fuel for extraneous political reasons not related to non-proliferation. They are also designed to remove a pretext for states proceeding with enrichment and reprocessing when, in fact, they have weapons purposes in mind.

There are several difficulties with these ideas, the main one being that they appear to be a 'solution in search of a problem'. It is the supplier states that have made all of the proposals, while the actual or potential importing countries are sceptical about them, claiming not to trust the supplier states to live up to their assurances and accusing them of wishing to preserve their technological superiority. A second argument made against such proposals, including by the companies that supply enriched uranium, is that the existing commercial market is sufficiently diversified to ensure supply. They point out that there has never been a case of a country having its supply cut off for political reasons. Even India, which was refused nuclear fuel for its CIRUS reactor after conducting its 1974 nuclear explosion, was able to receive it from France on the grounds of ensuring the 'safety' of the reactor (Perkovich 1999: 235).

Two of the proposals made for providing assurances of supply are relatively well advanced, one centred on the IAEA and one in Russia. Both schemes have run into surprisingly strong opposition in the Board from the very states that they are designed to assist; they see such projects as impinging on their Article IV rights under the NPT. The first, an IAEA Fuel Bank was proposed by the Nuclear Threat Initiative in September 2006. NTI committed $50 million to the IAEA to help it establish a stockpile of LEU to provide fuel assurances for non-proliferation-compliant states (NTI 2009a). Russia, meanwhile has proposed converting one of its existing nuclear facilities, located at Angarsk in Siberia, into an International Uranium Enrichment Centre (IUEC) to provide international enrichment services on a 'non-discriminatory' basis to facilitate assurances of supply. Currently, Russia owns 51 per cent of the project, which has already been launched, while Kazakhstan, Armenia and the Ukraine own 10 per cent each, leaving 19 per cent still available (Loukianova 2008). The Centre and its fuel will be under IAEA safeguards (Loukianova 2008) and the IAEA would have a role in deciding which states should have access to the fuel supply and in what circumstances. It is the first concrete outcome of a proposal in January 2006 by Russian President Vladimir Putin for a Global Power Infrastructure that would offer all countries equal access to nuclear energy, while ensuring non-proliferation objectives were met. The new IAEA Director General Yukiya Amano and the Secretariat now have the 'green light' to proceed with both proposals but with less than overwhelming support. However, with potential customers alienated from the proposal, it is not clear what their real future is.

The most radical solution to providing assurances of supply is the multilateralization of the entire nuclear fuel cycle. To prevent the unnecessary spread of enrichment and reprocessing technology many observers, including former IAEA Director General Mohamed ElBaradei, have proposed complete multilateralization (or internationalization) of the front and back ends of the nuclear fuel cycle (ElBaradei 2003b). Under such a regime, governments that currently have fuel cycle facilities would offer them up for multilateralization and all future plants would come under the same regime.

Views on the degree of internationalization needed are divided. Some argue that creating a few multilateral facilities to act as an incentive for states to forego indigenous enrichment and reprocessing is enough (NAS 2009: 52). Questions have also been raised about the extent of multilateral ownership of facilities, as well as about the national, multinational or international staffing of facilities and their locations (NAS 2009: 52). It is likely, however, that only a true multilateral regime, however radical an idea that may seem at present, will satisfy those states that are most concerned about discrimination and determined to preserve all of their Article IV rights. Such a regime would in any event have to be part of any serious move towards complete nuclear disarmament.

Tougher safeguards and other measures required to prevent a nuclear revival from increasing the risk of weapons proliferation will not be politically feasible unless the advanced nuclear energy states and those that have nuclear weapons (often one and the same) are prepared to forego options that they ask others to forego – the right to the full nuclear fuel cycle and the right to retain nuclear weapons in perpetuity. The existing nuclear energy states should thus commit soon to eventual international ownership and oversight of all enrichment, reprocessing and other sensitive fuel cycle facilities, with the modalities to be systematically studied in the coming years.

There is an obvious link between disarmament and calls for strengthening global governance in the realm of nuclear non-proliferation. Deepti Choubey, in a study for the Carnegie Endowment for International Peace, interviewed foreign ministry officials from 16 non-nuclear weapon states, including US allies both within and outside NATO. She concluded that 'The stark reality is that nuclear weapon states are in arrears and have a significant debt to pay before key non-nuclear weapon states will consider additional nonproliferation commitments' (Choubey 2008: 4).

The steps towards nuclear disarmament that the non-nuclear weapon states have long demanded is by now well known and encapsulated in several United Nations documents as well as the reports of various commissions and panels.[3] The most recent, the Independent Commission on Nuclear Nonproliferation and Disarmament (ICNND), released its report in December 2009. Commendably, the ICNND report for the first time seeks to link stages in the nuclear disarmament process not just to non-proliferation steps in general, but to 'progressive implementation of measures to reduce the proliferation risks associated with the expansion of civil nuclear energy' (ICNND 2009: 186). It also admirably seeks to establish linkages between disarmament measures and strengthening of the regimes for nuclear safety and security.

Encouragingly, there is a new wave of support and proposals for moving faster towards nuclear disarmament ('getting to zero'), including at least rhetorical support from the Obama administration. This may help break the deadlock between states arguing for ever tighter non-proliferation controls and those resisting on the grounds that the nuclear weapon states need to move faster to disarm as part of the non-proliferation grand bargain. The successful outcome of the NPT Review Conference in May 2010 augurs well, as does negotiation of the April 2010 New Strategic Arms Reduction Treaty between Russia and the United States that would lower deployed strategic nuclear weapons to around 1,550 each. US ratification of the CTBT, which could trigger Chinese ratification and a cascade towards entry into force, would clearly mark a major milestone long demanded by the non-nuclear weapon states. The commencement of negotiations on a Fissile Material Cut-Off Treaty (FMCT) in the Conference on Disarmament would be a further significant step. Such achievements would soften the harsh rhetoric and lower the political salience of moves to strengthen nuclear safeguards.

The integration of safety, security and non-proliferation

Although for the purposes of clarity this study has treated nuclear safety, nuclear security and nuclear non-proliferation separately, it is increasingly recognized that there is a strong relationship between them and that they have to be considered holistically if the global governance of all three is to be strengthened. Common principles, for instance, are seen to apply to safety and security, such as the philosophy of 'defence in depth'. Furthermore, safety and security measures designed to prevent unauthorized access to nuclear material can help prevent the acquisition of nuclear weapons by terrorists and other unauthorized entities. Again, non-proliferation measures, such as each country's State System of Accounting and Control, designed to help verify non-diversion of nuclear material to weapons purposes, also serve to deter unauthorized activities such as illicit trafficking and help the state account for and thus protect its nuclear assets.

As Richard Meserve points out with respect to nuclear power reactors, 'The massive structures of reinforced concrete and steel ... serve both safety and security objectives' (Meserve 2009: 107). A major breach of physical security, such as sabotage of a nuclear power plant, could pose serious safety risks. Meserve also notes that occasionally plant features and operational practices driven by safety considerations conflict with those that serve security purposes: 'Access controls imposed for security reasons can inhibit safety, limiting access for emergency response or egress in the event of a fire or explosion' (Meserve 2009: 107).

There is increasing recognition of the linkages and synergies. The '3-Ss' concept – safeguards, safety and security – was adopted by the 2008 Independent Commission of Eminent Persons convened to make recommendations on the role of the IAEA to 2020 and beyond (IAEA 2008a). It was later endorsed by the Group of 8 (G8) summit in Hokkaido, Japan in 2008 as a means of raising awareness of the importance of integrating the three fields and strengthening '3-S' infrastructure through international cooperation and assistance (G8 2008).

While the three regimes for safety, security and non-proliferation cannot for practical and political reasons be literally integrated, there needs to be heightened awareness of the linkages between them, both in order to achieve synergies and to avoid sensible actions being taken in one field that have deleterious consequences for others. Reactor design, the implementation of safeguards and other activities of the IAEA and national regulatory processes are just some of the areas that would be improved by pursuing such an integrated outlook. A nuclear energy revival should be exploited to ensure that guidance and assistance provided bilaterally and multilaterally to aspirant states in the areas of safety, security and safeguards is increasingly integrated, where feasible, and in all cases mutually supportive.

The future of the International Atomic Energy Agency

This study has also revealed the increasing centrality of the IAEA to the entire global nuclear governance realm. Always considered paramount in nuclear safeguards, the Agency has proved increasingly vital in nuclear safety, following Chernobyl, and in nuclear security, following 11 September 2001. States which have previously been lukewarm to the Agency in any of these areas need to recognize that, while not perfect, the IAEA has the greatest legitimacy, along with the highest levels of experience and capacity of any international body in the nuclear field. Considering that nothing short of international peace and security is at stake, the IAEA is a veritable security bargain.

It deserves increasing support along the following lines. Its budget, as recommended by the Commission of Eminent Persons on the Future of the Agency, should be doubled by 2020 to cope with the safety, security and safeguards demands arising from the increased use of nuclear energy. Proportionate increases should be made thereafter to maintain the purchasing power of the budget. In addition the IAEA should be funded to undertake a crash programme to upgrade its Seibersdorf facility to incorporate the latest technology and supportive infrastructure and to bring it up to the highest safety and security standards. The existing arrangements made for the Major Capital Investment Fund are inadequate and need to be boosted with the $50 million one-off injection of funds as recommended by former Director General ElBaradei. The IAEA Secretariat for its part needs to present the member states with a detailed plan on how the funds are to be spent rather than simply asking for a lump sum handout. The Agency should also be permitted to expand and renew its personnel resources, including by being exempted from constraining UN system rules.

In order that advice, guidance and assistance to aspirant states can be leveraged to produce the best possible outcomes for safety, security and non-proliferation the following should be considered:

- nuclear safety and security programmes should be funded from the IAEA regular budget rather than relying on voluntary contributions;
- Technical Cooperation specifically for new nuclear electricity programmes should be pegged to the recipient's safety, security and non-proliferation record and commitments; the Agency should cooperate closely with reactor supplier states and vendors in this endeavour; and
- the IAEA should be mandated to coordinate international assistance to new nuclear energy states aimed at improving their institutional capacities, especially legislative and regulatory ones, in advance of their acquisition of nuclear reactors; this would include assistance offered by other international bodies, governments and reactor vendors.

A global nuclear energy stakeholders forum?

While governments correctly retain the right to approve or reject the export of nuclear reactors or nuclear materials and other technologies by companies under their jurisdiction, industry cannot absolve itself of responsibilities by pretending that nuclear safety, security and non-proliferation are issues of 'high politics' that are entirely the province of governments. Industry has a strong self-interest in working more closely with the IAEA and other international bodies in ensuring that any nuclear revival does not rebound on its fortunes as a result of a serious accident, terrorist incident or nuclear weapons breakout.

This suggests that an international forum should be convened or an existing one adapted that brings together all states and companies involved in international nuclear power reactor sales in order to harmonize the criteria for proceeding with such sales. Such a forum could consider an industry code of conduct for nuclear reactor sales that restricts them to states which:

- are in full compliance with IAEA safeguards and an Additional Protocol;
- are party to all safety and security conventions;
- accept and implement high safety and security standards, including by participating in peer reviews;
- have established an appropriate national regulatory system;
- comply with UN Security Council Resolution 1540's reporting requirements; and
- voluntarily renounce sensitive nuclear technologies.

Additional considerations to be taken into account include: governmental stability, the quality of governance and regional security. Such a forum could emerge from the international activities of the former US-led Global Nuclear Energy Partnership or better still arise from an industry initiative designed to demonstrate good international citizenship and responsible business practice.

Conclusion

Global governance in the nuclear realm is already facing significant challenges even without the prospect of a nuclear energy revival. It is the obligation of the international community, governments, the nuclear industry and other stakeholders to do everything possible to ensure that a rise in the use of nuclear-generated electricity does not jeopardize current efforts being made to strengthen nuclear safety, security and non-proliferation. Indeed, the desire of states for the perceived benefits of nuclear energy should be levered to further reinforce of the various global governance arrangements.

The deal for aspiring states should be: if you want civilian nuclear power you have to agree to the highest international standards for avoiding accidents, terrorist seizure or attacks and diversion of materials to nuclear weapons. The deal for existing advanced nuclear states should be: if you want the newcomers to comply with a newly strengthened global regime that was not in place when you first acquired nuclear energy you have to multilateralize the fuel cycle and ultimately disarm yourselves of nuclear weapons.

There are formidable political obstacles to the achievement of such a deal. Many developing countries are increasingly suspicious of the intentions of the 'nuclear haves'. They interpret (or misinterpret) the NPT as granting them the 'inalienable' right to the peaceful uses of nuclear energy, often conveniently forgetting that even under the treaty this right is not unalloyed. They further seek to misinterpret this 'right' to mean that the existing nuclear energy states are obliged to provide unlimited and presumably discounted assistance to this end. These states, along with many developed non-nuclear weapon states, recall the false promises made in 1995 by the five recognized nuclear weapon states to move much faster towards nuclear disarmament in return for an indefinite extension of the NPT. They viewed with dismay the Bush administration's selective approach to its multilateral obligations under the treaty and have only been partially reassured by the Obama administration's policy turnaround.

For their part the nuclear 'haves' remain suspicious of the commitment of some non-nuclear weapon states to their continued non-nuclear status. The continuing saga of procrastination and prevarication by Iran over its nuclear intentions has poisoned the atmosphere at NPT review meetings and at the IAEA, while presenting foes of safeguards and nuclear governance generally with ammunition to support the contention that the system cannot be relied upon. While progress has been made by most of the nuclear weapon states to cut their nuclear arsenals, in the case of the United States and Russia dramatically, all of them continue to seek modernization of their nuclear forces and none talk of nuclear disarmament as anything but a distant goal to be achieved when the time is right. This implies that nuclear weapons are still regarded as useful, a point not lost on Iran, North Korea and others that may be contemplating nuclear 'hedging'. The position of the NPT outliers, India, Israel and Pakistan, complicates the picture further.

Whether the nuclear industry and governments favourably disposed to a nuclear energy expansion like it or not, this is the strategic and political context in which their plans will (or will not) be realized. This reinforces a major conclusion of this study that it is only enhanced involvement by all stakeholders, a holistic vision of global governance and ultimately some type of grand bargain, however fitfully arrived at, that will ensure a nuclear energy future that is both bountiful and innocuous.

Annexes

Annex 1: adherence to nuclear safety conventions by SENES states

State	Convention on nuclear safety	Convention on early notification of a nuclear accident	Convention on assistance in the case of a nuclear accident or radiological emergency	Joint convention on the safety of spent fuel management and radioactive waste management	Vienna convention on civil liability for nuclear damage
Albania	Unsigned	In force	In force	Unsigned	Unsigned
Algeria	Signed	In force	In force	Unsigned	Unsigned
Bahrain	Unsigned	Unsigned	Unsigned	Unsigned	Unsigned
Bangladesh	In force	In force	In force	Unsigned	Unsigned
Belarus	In force	In force	In force	Unsigned	Unsigned
Egypt	Signed	In force	In force	Unsigned	Unsigned
Ghana	In force	In force	In force	Unsigned	Unsigned
Indonesia	In force	In force	In force	Unsigned	Unsigned
Iran	Unsigned	In force	In force	Unsigned	Unsigned
Israel	Signed	In force	In force	Unsigned	Signed
Italy*	In force	In force	In force	In force	Unsigned
Jordan	In force	In force	In force	Unsigned	Unsigned
Kazakhstan	Signed	In force	In force	Signed	Unsigned
Kenya	Unsigned	Unsigned	Unsigned	Unsigned	Unsigned
Kuwait	In force	In force	In force	Unsigned	Unsigned
Libya	In force	In force	In force	Unsigned	Unsigned
Malaysia	Unsigned	In force	In force	Unsigned	Unsigned
Mongolia	In force	In force	In force	Unsigned	Unsigned
Morocco	Signed	In force	In force	Unsigned	Signed
Namibia	Unsigned	Unsigned	Unsigned	Unsigned	Unsigned
Nigeria	In force	In force	In force	Unsigned	Unsigned
Oman	Unsigned	Unsigned	Unsigned	Unsigned	Unsigned
Phillippines	Signed	In force	In force	Signed	Unsigned
Poland	In force	In force	In force	In force	In force
Qatar	Unsigned	Unsigned	Unsigned	Unsigned	Unsigned
Saudi Arabia	Unsigned	Unsigned	Unsigned	Unsigned	Unsigned
Senegal	In force	In force	In force	In force	In force
Syria	Signed	Signed	Signed	Unsigned	Unsigned
Thailand	Unsigned	In force	In force	Unsigned	Unsigned
Tunisia	Signed	In force	In force	Unsigned	Unsigned
Turkey	In force	In force	In force	Unsigned	Unsigned
United Arab Emirates	In force	In force	In force	Unsigned	In force
Venezuela	Unsigned	Unsigned	Unsigned	Unsigned	Unsigned
Vietnam	Unsigned	In force	In force	Unsigned	Unsigned

Legend: Unsigned ☐ Signed ▨ In force ▇

Source: IAEA (2009i). Available at: www.iaea.org/Publications/Documents/Conventions/index.html (accessed 24 November 2009).

Note
* Italy has ratified the 1960 Paris Convention on Nuclear Liability and the Joint Protocol Relating to the Application of the Vienna Convention and the Paris Convention, thereby obliging it to be bound by the constraints of the 1963 Vienna Convention.

Annex 2: adherence to nuclear security conventions by SENES states

State	Convention on the physical protection of nuclear material	Amendment to the convention on the physical protection of nuclear material	International convention for the suppression of acts of nuclear terrorism
Albania	In force	Unsigned	Signed
Algeria	In force	Unsigned	In force
Bahrain	Unsigned	In force	In force
Bangladesh	In force	Unsigned	In force
Belarus	In force	Unsigned	In force
Egypt	Unsigned	Unsigned	Signed
Ghana	In force	Unsigned	Signed
Indonesia	In force	Unsigned	Unsigned
Iran	Unsigned	Unsigned	Unsigned
Israel	In force	Unsigned	Signed
Italy	In force	Unsigned	Signed
Jordan	In force	In force	Signed
Kazakhstan	In force	Unsigned	In force
Kenya	In force	Unsigned	In force
Kuwait	In force	Unsigned	Signed
Libya	In force	In force	In force
Malaysia	Unsigned	Unsigned	In force
Mongolia	In force	Unsigned	In force
Morocco	In force	Unsigned	Signed
Namibia	In force	Unsigned	Unsigned
Nigeria	In force	In force	Unsigned
Oman	In force	Unsigned	Unsigned
Phillippines	In force	Unsigned	Signed
Poland	In force	In force	In force
Qatar	In force	Unsigned	Signed
Saudi Arabia	In force	Unsigned	In force
Senegal	In force	Unsigned	Signed
Syria	Signed	Unsigned	Signed
Thailand	Unsigned	Unsigned	Signed
Tunisia	In force	Unsigned	Unsigned
Turkey	In force	Unsigned	Signed
United Arab Emirates	In force	Unsigned	In force
Venezuela	Unsigned	Unsigned	Unsigned
Vietnam	Unsigned	Unsigned	Unsigned

Legend: Unsigned ☐   Signed ▨   In force ■

Source: IAEA (2009i). Available at: www.iaea.org/Publications/Documents/Conventions/index.html (accessed 21 December 2009). See also 'International Convention for the Suppression of Acts of Nuclear Terrorism: Status', United Nations Treaty Collection, United Nations, http://treaties.un.org/Pages/DB.aspx?path=DB/MTDSGStatus/pageIntro_en.xml (accessed 21 December 2009).

Annex 3: adherence to nuclear safeguards by SENES states

State	Comprehensive safeguards agreement	Additional protocol	Integrated safeguards	Small quantities protocol	
				Old	*New*
Albania	In force				
Algeria	In force	Signed			
Bahrain	In force	In force			In force
Bangladesh	In force	Signed	In force		
Belarus	In force	Signed			
Egypt	In force				
Ghana	In force	In force	In force		
Indonesia	In force				
Iran	In force	Signed			
Israel					
Italy	In force				
Jordan	In force	In force		In force	
Kazakhstan	In force				
Kenya	In force				In force
Kuwait	In force	In force		In force	
Libya	In force	In force			
Malaysia	In force	Signed			
Mongolia	In force	In force		In force	
Morocco	In force	Signed		In force	
Namibia	In force	Signed		In force	
Nigeria	In force	In force			
Oman	In force			In force	
Phillippines	In force		In force		
Poland	In force		In force		
Qatar	In force				In force
Saudi Arabia	In force			In force	
Senegal	In force	Signed		In force	
Syria	In force				
Thailand	In force	Signed			
Tunisia	In force	Signed			
Turkey	In force	In force			
United Arab Emirates	In force			In force	
Venezuela	In force				
Vietnam	In force	Signed			

Legend: Unsigned ☐   Signed ▨   In force ■

Sources: IAEA (2008b and 2009c).

Annex 4: proposals for assurances of supply and/or multilateralization of the nuclear fuel cycle

US Proposal on a Reserve of Nuclear Fuel (2005) In September 2005, the United States announced that it would commit 17 metric tons of HEU to be down-blended to LEU to act as a reliable supply of fuel for states that forego enrichment and reprocessing.

US Global Nuclear Energy Partnership (2006) As a part of GNEP the United States proposed that a consortium of countries ensure reliable access to fuel for countries that forego enrichment and reprocessing (Pomper 2010).

World Nuclear Association Proposal (2006) In May 2006 the WNA Working Group on Security of the International Fuel Cycle proposed a three-stage assurances of supply arrangement:

1   basic supply security provided by the existing world market;
2   collective guarantees by enrichers supported by governmental and IAEA commitments;
3   government stocks of enriched uranium product (EUP) (WNA 2006).

The WNA proposal emphasizes that the existing sources of supply are sufficient to ensure reliable supply for all states.

Concept for a Multilateral Mechanism for Reliable Access to Nuclear Fuel The six enrichment supplier states – France, Germany, the Netherlands, Russia, the UK and the United States – proposed two levels of assurance of supply of nuclear fuel. The first level, 'basic assurances', commits suppliers to substitute for each other in cases of interruption of supply. The second level, 'reserves', refers to a virtual or physical supply of fuel set aside if basic assurances are to fail.

IAEA Standby Arrangements System Japan proposed that an information system be established to complement the six enrichment supplier states' proposal. The system would track national fuel cycle capacities throughout the entire fuel cycle so that other states always have a clear view of available supply.

Nuclear Fuel Assurance Proposal This proposal by the UK, previously named the Enrichment Bonds Proposal, proposes offering enrichment bonds that guarantee that national suppliers would not be prevented from supplying enrichment services, and to provide prior consent for export assurances.

Multilateral Enrichment Sanctuary Project In May 2007 Germany proposed a multilateral uranium enrichment centre with extra-territorial status that would operate under IAEA control on a commercial basis. This proposal has since evolved into a Multilateral Enrichment Sanctuary Project (MESP) which would be supervised by the IAEA, but owned and operated by a multinational consortium.

Multilateralization of the Nuclear Fuel Cycle Austria proposed a two-track system, the first involving increasing transparency beyond current IAEA safeguards, the second creating a nuclear fuel bank to act as a hub for all nuclear fuel transactions.

Nuclear Fuel Cycle non-paper In June 2007 the European Union, in response to many of the above proposals, emphasized in a non-paper the importance of maintaining flexibility in considering approaches to the fuel cycle. The non-paper proposed proliferation resistance, assurance of supply, consistency with equal rights and obligations and market neutrality as important criteria for any fuel cycle proposal.

International Nuclear Fuel Agency In a draft paper released in 2009, Thomas Cochran and Christopher Paine proposed that a new agency, the International Nuclear Fuel Agency (INFA), be established to certify design, construction and operation of all uranium enrichment facilities worldwide (Cochran and Paine 2009). All enrichment activities would be conducted inside long-term 'Sovereign Secure Leased Areas' that INFA would lease from governments; enrichment facilities would still be under national or private ownership but INFA would ensure that safeguards were maintained and would control entry and egress from the site as well as end use of the enriched fuel supplied (Cochran and Paine 2009).

(Source: adapted from Yudin 2009)

Notes

Introduction

1 The first use of the term 'renaissance' appears to have occurred as early as 1985 in an article by Alvin M. Weinberg, Irving Spiewak, Doan L. Phung and Robert S. Livingston, four physicists from the Institute for Energy Analysis at Oak Ridge, Tennessee, in the journal *Energy* entitled 'A Second Nuclear Era: A Nuclear Renaissance'. The first use of the term in the new millennium appears to have been in an article in *Power* magazine, volume 144, number 3, of 1 May 2000, entitled 'Nuclear Power Embarks on a Renaissance'.
2 Book examples include: Cravens (2007), Herbst and Hopley (2007), Nuttall (2005) and Tucker (2008).
3 See *Guide to Global Nuclear Governance: Safety, Security and Nonproliferation* (Alger 2008).

1 Assessing a nuclear energy revival: the drivers

1 A study by Pacific Northwest Laboratory (PNNL) found that there is already enough generating capacity to replace as much as 73 per cent of the US conventional fleet with electric cars if charging was managed carefully using 'smart grid' off-peak electricity in the evenings.
2 Actinides are elements on the periodic table with atomic numbers ranging from 89 to 103. All actinides are radioactive and only two – thorium and uranium – occur naturally. In the nuclear energy industry, the term actinides often refers to just the long-lived, highly radioactive materials contained in spent nuclear fuel, such as neptunium and americium. Due to their radioactivity, actinides can have a variety of negative effects on human health and the environment depending on their availability and exposure to them.
3 Lovelock has calculated that the maximum amount of carbon in the atmosphere consonant with the planet 'on which civilization developed and to which life is adapted' is 350 parts per million CO_2. The current level is 387.
4 The technology focus of the Nuclear Power 2010 programme is Generation III+ advanced light water reactor designs which offer advancements in safety and economics over the Generation III designs certified by the Nuclear Regulatory Commission (NRC) in the 1990s.
5 The IAEA has a wealth of information on SMRs at: www.iaea.org/NuclearPower/SMR/.
6 For details of additional SMR research efforts directed at 'near term deployment', see NEA (2008: 381–382).
7 Currently Argentina, Brazil, Canada, Euratom, France, Japan, China, South Korea, South Africa, Russia, Switzerland, the UK and the United States. The NEA hosts the Technical Secretariat.

8 Argentina, Armenia, Belarus, Belgium, Brazil, Bulgaria, Canada, Chile, China, Czech Republic, France, Germany, India, Indonesia, Japan, South Korea, Morocco, Netherlands, Pakistan, Russia, Slovakia, South Africa, Spain, Switzerland, Turkey, Ukraine, the United States and the European Commission.
9 With Algeria, Brazil, India, Jordan, Libya, Morocco, Qatar, Russia, Tunisia, the UAE, UK and Vietnam.
10 The OECD has 30 members: Australia, Austria, Belgium, Canada, Czech Republic, Denmark, Finland, France, Germany, Greece, Hungary, Iceland, Ireland, Italy, Japan, Korea, Luxembourg, Mexico, the Netherlands, New Zealand, Norway, Poland, Portugal, Slovak Republic, Spain, Sweden, Switzerland, Turkey, United Kingdom and the United States. The only two OECD members not part of the NEA are New Zealand and Poland.

2 Assessing a nuclear energy revival: the constraints

1 This included £500 million interest during construction plus on-site waste storage costs.
2 The equivalent of $5,000-$6,000/kW (Moody's Corporate Finance (2007). Quoted in Squassoni (2009b: 30)).
3 See chart showing comparisons in Smith (2006: 38).
4 See chart showing comparisons in Smith (2006: 38); also NEA (2005: 36).
5 As the US Congressional Budget Office (CBO) notes,

> If the levelized cost of a technology exceeded anticipated prices for electricity, merchant generators would be unlikely to invest in new capacity based on that technology because the expected return would not justify the amount of risk they would have to incur. State utility commissions commonly direct regulated utilities to meet anticipated demand for new capacity using the technology with the lowest levelized cost.

6 Business schools teach that a decision about the financial viability of a project should not be confused with a decision to invest, since the former is concerned with whether the project is likely to turn a profit, whereas the latter is concerned with comparative risk and rates of return on investment.
7 As the IEA notes, however, there does not have to be uniform incentives with the same value for all technologies. It argues for subsidies for the more expensive alternatives.
8 See Schewe (2007: Chapter 8).
9 The increase was due in part to changes in political and public views of nuclear energy following the Chernobyl accident, with subsequent alterations in the regulatory requirements.
10 While, as Mycle Schneider points out, France's high level of reactor standardization has multiple technical and economic advantages, it has also led to systematic multiplication of problems in the reactor fleet.
11 For an example, see a programme called the business risk management framework (BRMF).
12 See chart in CBO (2008: 26).
13 For details see Table 1.1, Incentives provided by the Energy Policy Act of 2005, CBO Report: 11. The tax credit provides up to $18 in tax relief per megawatt hour of electricity produced at qualifying power plants during the first eight years of operation. The average wholesale price of electricity in the United States in 2005 was about $50 per megawatt. The loan programme provides a federal guarantee on debt covering as much as 80 per cent of construction costs but it also applies to innovative fossil fuel or renewable technologies (CBO 2008: 8–9).
14 According to Paul Brown, the UK taxpayer has already underwritten all the debts and liabilities of British Energy so the company can never go bankrupt (Brown 2008: 3).

15 A Nuclear Transparency and Safety Act was passed by the National Assembly only in 2006 (Law no. 2006–686, 13 June 2006). This apparently has had limited effects in making financial details available (Schneider 2008b: 6–7).

16 The 2009 delivery European Union Allowances (EUAs) closed in trading on 18 March 2009, on the European Climate Exchange at €12.50/metric ton (mt). According to Deutsche Bank carbon analyst Mark Lewis, if the value reaches €35/mt between 2013 and 2020 as some predict, nuclear power could become the cheapest form of new electricity in the EU. At that price coal and gas would cost €86/MWh; CSS capacity €102/MWh; and nuclear only €60/MWh. According to *The Economist* 'power industry bosses' believe the price of carbon would need to reach €50 for nuclear to become economic (*The Economist*, 14 November 2009: 67).

17 While reactor vendors prefer large forgings to be in a single piece, it is possible to use split forgings welded together, but these need continued checking throughout the plant's lifetime.

18 WNU is a non-profit corporation and public–private partnership, pursuing an educational and leadership-building mission through programmes organized by the WNU Coordinating Centre (WNUCC) in London. The WNUCC's secretariat is composed mainly of nuclear professionals seconded by governments; the IAEA further assists with financial support for certain activities. The nuclear industry provides administrative, logistical and financial support (WNU 2010).

19 For a comprehensive history of the 1982 Nuclear Waste Policy Act and proposed Yucca Mountain repository see Vandenbosch and Vandenbosch (2007).

20 See Table 3.6 in Bratt (2006: 79).

3 Assessing the likelihood of a revival

1 GWe or GW(e) is a measure of electrical energy. 1 GWe = 1,000 watts of electricity (MWe) or 109 watts. GWth is a measure of thermal energy.

2 The projections were higher than in 2007 and 2008, when the agency predicted 691 and 748 GWe respectively by 2030.

3 The IAEA does not project reactor numbers.

4 In 1974 the IAEA issued the last of its completely optimistic forecasts of global nuclear capacity, predicting that the existing figure of about 55,000 MWe would multiply more than ten-fold by 1985 to almost 600,000 MWe. By 1990 there would supposedly be more than one million and by the year 2000 almost three million.

5 The 32 MW reactors are the Akademik Lomonosov 1 and 2, KLT-40S Floati. The other seven are 750–1,085 MW each (NEA 2008: 424).

6 Work originally began on the two Bushehr reactors in 1974 by the German firms Siemens and its subsidiary Kraftwerke Union, but stopped when the Shah was overthrown in 1979. The site was bombed during the 1980–1988 Iran/Iraq war. The Russians were contracted to complete one of the reactors in 1995 which is likely to come online in 2011 (Bahgat 2007: 20–22).

7 The 2003 MIT study recommended a focus on such light water reactors and 'some R&D' on the high temperature gas reactor (HTGR) because of its potential for greater safety and efficiency of operation.

8 For a series of national case studies on Canada, China, France, India, Russia, the UK and the United States that were commissioned as part of the CIGI/CCTC Nuclear Energy Futures Project see www.cigionline.org.

9 For a complete study of Canada's situation see Cadham (2009).

10 The Canadian government announced in June 2009 that it intended to take Canada out of the radioisotope business, following the latest breakdown in the world's oldest nuclear reactor at Chalk River, previously the supplier of more than 50 per cent of the world's radionuclides market, and the cancellation in 2007 of the two intended

replacement reactors, the AECL-designed Maple 1 and 2, due to insurmountable technical difficulties.

11 At the time of writing, Areva was seeking to sell its electricity transmission and distribution business.

12 For details of improvements see Uranium Institute, 'Post-accident changes', reproduced in Steed (2007: 271–274).

13 In December 2008, South African utility Eskom cancelled its tender for a turnkey nuclear power station, saying the magnitude of the investment was too much for it to handle. In June 2007 Eskom had announced plans for up to 20,000 MW of new nuclear power by 2025. Areva and Westinghouse had both bid for two new power stations. The reported estimated cost of $9 billion had escalated to $11 billion with the devaluation of the Rand.

14 Continuously updated summaries of each SENES state's progress are available at www.cigionline.ca/senes.

15 Grid capacity and electricity production are not interchangeable since different factors are involved in calculating them. Nevertheless, because there is a strong correlation between the two, electricity production is a suitable proxy for the size of a country's energy infrastructure.

4 The current status of global nuclear governance – the nuclear safety regime

1 See discussion of the safety of nuclear reactors in theory in Trevor Findlay, 'Nuclear Safety', *The Future of Nuclear Energy to 2030 and its Implications for Safety, Security and Nonproliferation*, Centre for International Governance Innovation (CIGI), Waterloo, www.cigionline.org/publications/2010/2/future-nuclear-energy-2030.

2 See chart in Smith (2006: 172).

3 For a brief description of incidents in the civilian nuclear industry to 1998 see Ramsey and Modarres (1998: 105–136).

4 Canada, France, Germany, Italy, Japan, the United Kingdom and the United States.

5 The reports that are accessible are impressively detailed. Five states – Canada, Luxembourg, Slovenia, Switzerland and the UK – also included their detailed responses to questions from other parties.

6 France was an exception, but has recently taken steps to improve. In December 2007 its Institute of Radiological Protection and Nuclear Safety (IRSN) for the first time publicly released one of its reports, on the safety of Electricité de France's reactor fleet in 2007 (IRSN 2008).

7 Safety Guides are submitted by the Director General to the IAEA Publications Committee.

8 While nuclear safety is concerned with ensuring the safe operation of nuclear facilities and other activities, radiological standards and protection are designed to shield the public, workers and the environment from the harmful effects of radiation.

9 These include a Safety Reports Series, a Safety Series, a Services Series, a Technical Documents Series, a Radiological Assessment Reports Series and the International Nuclear Safety Advisory Group Series. The IAEA also issues the Provisional Safety Standards Series, the Training Course Series, the IAEA Services Series, a Computer Manual Series, Practical Radiation Safety Manuals and Practical Radiation Technical Manuals and Handbook on Nuclear Law.

10 Private communication with IAEA official, October 2008.

11 Private communication with WANO, December 2009. WANO uses the term 'station' to include sites with one or more reactor, depending on how their regional centres handle each review.

12 Canada Deuterium Uranium reactor.

13 Water-cooled, water-moderated reactor.

14 Spent fuel is nuclear fuel that has been irradiated in a reactor core. It may be reprocessed to produce uranium and plutonium which may be recycled as reactor fuel. Radioactive waste is defined as radioactive material in gaseous, liquid or solid form for which there is no foreseen further use and which has been declared as radioactive waste.

15 For information on the negotiation history see Friedrich and Finucane (2001).

16 The following paragraph adapted from Gonzáles (2002: 290–291).

17 The following paragraph adapted from the IAEA (2003b: 91).

18 Among the contracting parties are the World Health Organization (WHO), the World Meteorological Organization (WMO), the Food and Agricultural Organization (FAO) and Euratom.

19 Among the contracting parties are the WHO, the WMO, FAO and Euratom.

20 The members, in addition to the IAEA, include: the European Commission, the European Police Office (EUROPOL), the Food and Agriculture Organization of the United Nations (FAO), the International Civil Aviation Organization (ICAO), the International Maritime Organization (IMO), the United Nations Scientific Committee on the Effects of Atomic Radiation (UNSCEAR), the International Criminal Police Organization (INTERPOL), the Organisation for Economic Co-operation and Development (OCED)/Nuclear Energy Agency (NEA), the Pan American Health Organization (PAHO), the United Nations Environment Programme (UNEP), the United Nations Office for the Co-ordination of Humanitarian Affairs (UN/OCHA), the United Nations Office for Outer Space Affairs (UN/OOSA), the World Health Organization (WHO) and the World Meteorological Organization (WMO).

21 In the case of the US, according to the Congressional Budget Office, the subsidy probably amounts to less than 1 per cent of the levelized cost of new nuclear capacity (cited in Kidd 2009b).

22 There are other examples of international law relating to trans-boundary liability, such as the 1969 Civil Convention on Oil Pollution Damage or the 1972 Convention on Damage Caused by Space Objects.

23 The third such Regional Workshop on Liability for Nuclear Damage was held in South Africa in February 2008. The fourth was held early in Abu Dhabi in the UAE in December 2009 for countries that have expressed an interest in launching a nuclear power programme (IAEA 2009e: 12; IAEA 2009o).

24 See also Tetley (2006).

5 The current status of global nuclear governance – nuclear security and non-proliferation

1 See Allison (2005) and Charles D. Ferguson and William C. Potter with Amy Sands, Leonard S. Spector and Fred L. Wehling, *The Four Faces of Nuclear Terrorism*, Center for Nonproliferation Studies, Monterey, CA, 2004. For a sceptical view see Mueller (2010).

2 The 1985 Treaty of Rarotonga, which created a nuclear weapon-free zone in the South Pacific, bans nuclear dumping but does not concern itself with nuclear safety or security.

3 The resolution has been extended twice, in 2006 (Resolution 1673) for two years and in 2008 (Resolution 1810) until 2011.

4 The implementing guides most relevant to the security of civilian nuclear facilities are: Engineering Safety Aspects of the Protection of Nuclear Power Plants against Sabotage; Nuclear Security Culture; Preventive and Protective Measures against Insider Threats; Security in the Transport of Radioactive Material; Development, Use and Maintenance of the Design Basis Threat.

5 For example, see Goldschmidt (2007 and 2009: 4).

6 In addition Mongolia declared itself a nuclear weapon-free zone in 1992 (it entered into force in 2000). The 1950 Antarctic Treaty also forbids the deployment of nuclear weapons in the Antarctic. The 1967 Outer Space Treaty prohibits the stationing of

nuclear weapons in outer space, or on the moon or other celestial bodies, while the 1971 Seabed Treaty prohibits the emplacement of weapons on the seabed.

7 Australia, Belgium, Brazil, Canada, Czechoslovakia, France, India, Portugal, South Africa, the Soviet Union, the United Kingdom and the United States (Fischer 1997: 39–40).

8 Encapsulated in document INFCIRC/153 (IAEA 1972).

9 The quantities are 8 kg of plutonium and uranium-233, 25 kg of uranium-235 enriched to 20 per cent or more, 75 kg of uranium-235 enriched to less than 20 per cent, 10 tons of natural uranium and 20 tons of natural uranium or thorium (IAEA 1998: 53).

10 For Romania see Mozley (1998); for Egypt and South Korea see US GAO (2005: 20); for Taiwan see Quester (1985); IAEA (2009c); and IAEA (2008b: 68).

11 NTI is a euphemism for all sources of information available to an individual state for monitoring the behaviour of other states, including treaty compliance. For further details, see UNIDIR (2003: 20–22).

12 Article XII.5.

13 For a thorough critique of IAEA nuclear safeguards see Sokolski (2007).

14 For a series of papers analysing the IAEA, see Sokolski (2008).

15 The United States has concluded nine such agreements, including with Liberia and Panama, the two largest ship registrars (US Department of State 2009c).

6 Implications of a nuclear revival for global governance

1 US officials are currently concerned about a new recruit to Al Qaeda in Yemen who, although he had a low security clearance, was previously a maintenance worker at five nuclear plans along the US East Coast (Faddis 2010).

2 The computer analyses, which cost more than $1 million, are summarized in EPRI (2002).

3 Presentation by Thomas B. Cochran, Senior Scientist, Nuclear Program, Natural Resources Defense Council to the conference on 'Next Generation Nuclear Security: Meeting the Global Challenge', organized by the Fissile Materials Working Group, 12 April 2010, Washington, DC.

4 The following section draws significantly on research conducted by Justin Alger, Researcher and Administrator at Canadian Centre for Treaty Compliance (CCTC) at the Norman Paterson School of International Affairs, Carleton University, Ottawa, Canada for the Nuclear Energy Futures Project run by the Centre for International Governance Innovation (CIGI) in Waterloo, Ontario, Canada in cooperation with the CCTC. See Alger (2009).

5 There is considerable overlap between the basic scientific disciplines required for a nuclear energy programme and a nuclear weapons programme. Such disciplines including: nuclear, chemical, metallurgical and electrical engineering, physics, mathematics, computer science and chemistry. But these also overlap with many other industrial and technological enterprises. Examples of peaceful/military crossover in nuclear engineering include fissile atom depletion and production calculations, criticality calculations and nuclear reactor design (US GAO 1979). Some of the disciplinary overlap – particularly in chemical engineering – relates to sensitive fuel cycle technologies such a enrichment and reprocessing.

6 A US Congressional Office of Technology Assessment concluded in 1995 that

> In general, assistance at the level and for the purposes provided by the IAEA makes little direct contribution to a nuclear weapon program. However, the skills and expertise that might be acquired by a state through such assistance could be relevant, both in terms of basic knowledge in dealing with nuclear materials and nuclear technology, and also possibly in terms of extrapolating techniques a state first learns through IAEA technical assistance.
>
> (OTA 1995: 54)

7 For detailed historical accounts of nuclear assistance to India, Israel, North Korea, Pakistan and South Africa, see the *Nuclear Threat Initiative's* country profiles. Available at: www.nti.org/e_research/profiles/index.html.

8 Most recently by the ICNND; see ICNND (2009: 126).

9 For an account of the CANDU controversy, see Bratt (2006: 46–47).

10 Pakistan had help from France in building the Chasma and Pinstech reprocessing plants, China is suspected of helping Pakistan with the Kahuta enrichment plant. France also assisted Israel to construct the Dimona reprocessing plant. North Korea received reprocessing technology from the Soviet Union in the 1960s, and is suspected of receiving designs and components for an enrichment plant supplied by Pakistan. For further details, see Kroenig (2009) and NTI (2009b).

11 For more information on India's nuclear programme see Perkovich (1999).

12 During the 1960s and 1970s the US government allowed a few dozen foreign scientists to be involved in unclassified research relating to enrichment or reprocessing. A declassified 1979 report by the US Comptroller General noted that:

> Department of Energy officials said that sensitive areas of nuclear technology have been examined and precautions have been taken, but it is difficult to draw a firm line between what is and is not sensitive; it is a matter of degree.
>
> (US GAO 1979: i–vi)

13 For more information about the challenges of weaponization see NTI (2009a).

14 The major regions not covered by nuclear weapon-free zones are Europe, East Asia, South Asia and the Middle East.

15 'At what cost, success', MANNET (Management Network), Chambésy, Switzerland, 14 October 2002, cited in IAEA (2008c: 29).

Conclusions and recommendations

1 Undersecretary of State for Arms Control and International Security John Bolton told reporters that the United States would not 'apply a double standard' to South Korea (Kerr 2004).

2 For an examination of technological and other possibilities to 2050 and beyond, see Feiveson *et al.* (2008).

3 Among these are the Final Document of the First United Nations Special Session on Disarmament (UNSSOD 1) in 1978 (UN 1978); the 1996 Canberra Commission on the Elimination of Nuclear Weapons (Canberra Commission 1996); the Thirteen Practical Steps agreed at the 2000 NPT Review Conference; and the 2006 Weapons of Mass Destruction Commission (WMD Commission 2006).

References

1540 Committee (2010). 'List of National Reports By Submitting Member States', United Nations. Available at: www.un.org/sc/1540/nationalreports.shtml.

Accenture Newsroom (2009). 'Consumers Warm to Nuclear Power in Fight against Fossil Fuel Dependency', 17 March. Available at: http://newsroom.accenture.com/article_display.cfm?article_id=4810.

Acton, James (2009). 'The Myth of Proliferation-resistant Technology', *Bulletin of the Atomic Scientists*, November/December.

Acton, James and Carter Newman (2006). 'IAEA Verification of Military Research and Development', *Verification Matters*, No. 5, Verification Research, Training and Information Centre (VERTIC), London.

Albright, David (2007). 'Nuclear Black Markets: Pakistan, A.Q. Khan, and the Rise of Proliferation Networks', Institute for Science and International Security (ISIS), Written Testimony before the House Committee on Foreign Affairs' Subcommittee on the Middle East and South Asia and the Subcommittee on Terrorism, Nonproliferation, and Trade, 27 June. Available at: www.globalsecurity.org/wmd/library/congress/2007_h/070627-albright.htm.

Albright, David and Lauren Barbour (2000). 'Status Report: Civil Plutonium Transparency and the Plutonium Management Guidelines', Institute for Science and International Security (ISIS), Washington, DC, 1 January. Available at: http://isis-online.org/isis-reports/detail/status-report-civil-plutonium-transparency-and-the-plutonium-management-gui/17.

Alexander, Ryan (2009). 'The Real Cost of New U.S. Nuclear Reactors', *Bulletin of the Atomic Scientists*, 21 August. Available at: www.thebulletin.org/web-edition/op-eds/the-real-cost-of-new-us-nuclear-reactors.

Alger, Justin (2008). 'A Guide to Global Nuclear Governance: Safety, Security and Nonproliferation', *Nuclear Energy Futures Special Publication*, September. Available at: www.cigionline.org/publications/2008/9/guide-global-nuclear-governance-safety-security-and-nonproliferation.

Alger, Justin (2009). 'From Nuclear Energy to the Bomb: the Proliferation Potential of New Nuclear Energy Programs', *Nuclear Energy Futures Paper*, No. 6, September. Available at: www.cigionline.org/publications/2009/9/nuclear-energy-bomb-proliferation-potential-new-nuclear-energy-programs.

Allison, Graham (2004). *Nuclear Terrorism: The Ultimate Preventable Catastrophe*. New York: Owl Books.

America.gov (2010a). Communiqué from Washington Nuclear Security Summit, 13 April. Available at: www.america.gov/st/texttrans-english/2010/April/20100413171855eaifas0.6155773.html.

America.gov (2010b). Washington Nuclear Security Summit Work Plan, 13 April. Available at: www.america.gov/st/texttrans-english/2010/April/20100413182810ihecuor0.8 188702.html.

Areva (2009). 'Operations', Areva Group, Paris. Available at: www.areva.com/servlet/operations-en.html.

Asian Development Bank (ADB) (2009). 'Energy Policy – June 2009', Policy Paper, June.

ATMEA (2009). 'ATMEA1: The IAEA Completes Reactor Safety Features Review', ATMEA, 7 July. Available at: www.atmea-sas.com/scripts/ATMEA/publigen/content/templates/Show.asp?P=247&L=EN.

Bahgat, Gawdat (2007). *Proliferation of Nuclear Weapons in the Middle East*. Gainesville: University Press of Florida.

Bajaj, Vikas (2010). 'India's Woes Reflected in Bid to Restart Old Plant', *New York Times*, 22 March. Available at: www.nytimes.com/2020/03/23/business/global/23enron. html.

BAS (2009). 'Interview: Mohamed ElBaradei', *The Bulletin of the Atomic Scientists*, September/October, pp. 1–9.

BBC News (2009). 'Obama Promotes Nuclear-free World', 5 April. Available at: http://news.bbc.co.uk/2/hi/europe/7983963.stm.

Bergenäs, Johan (2009). 'Sweden Reverses Nuclear Phase-Out Policy', Center for Nonproliferation Studies/Nuclear Threat Initiative Issue Brief, Monterey Institute for International Studies, November. Available at: www.nti.org/e_research/e3_sweden_reverses_nuclear_phaseout_policy.html.

Birtles, Peter (2009). 'Time to Forge Ahead', *Nuclear Engineering International*, March.

Boese, Wade (2005). 'Key U.S. Interdiction Initiative Claim Misrepresented', *Arms Control Today*, 35, July/August. Available at: www.armscontrol.org/act/2005_07–08/Interdiction_Misrepresented.

Boone, David (2009). 'In at the Ground Floor', *Nuclear Engineering International*, July, pp. 8–9.

Borger, Julian (2007). 'Nuclear Watchdog Might Not Cope in Atomic Crisis', *Guardian*, 22 June. Available at: www.guardian.co.uk/world/2007/jun/22/northkorea.

Bradford, Peter (2009). 'The Nuclear Renaissance Meets Economic Reality', *Bulletin of the Atomic Scientists*, Vol. 65, No. 6 (November/December), pp. 60–64.

Bradsher, Keith (2009). 'Nuclear Power Expansion in China Stirs Concerns', *New York Times*, 16 December. Available at: www.nytimes.com/2009/12/16/business/global/16 chinanuke.html.

Bratt, Duane (2006). *The Politics of CANDU Exports, IPAC Series in Public Management and Governance*. Toronto: University of Toronto Press.

Brett, Patricia (2009). 'The Dilemma of Aging Nuclear Plants', *New York Times*, 20 October. Available at: www.nytimes.com/2009/10/20/business/global/20renuke.html.

British Petroleum (BP) (2009). '"Electricity Generation" and "Nuclear Energy – Consumption"', in *Statistical Review of World Energy 2009: Historical Data*, British Petroleum, June. Available at: www.bp.com/productlanding.do?categoryId=6929&contentId=7044622.

Broodryk, Amelia and Noël Stott (2009). 'Africa Is Now Officially a Zone Free of Nuclear Weapons', Arms Management Programme, Pretoria: Institute for Security Studies, 12 August. Available at: www.iss.co.za/index.php?link_id=5&slink_id=8113 &link_type=12&slink_type=12&tmpl_id=3.

Brown, Paul (2008). 'Voodoo Economics and the Doomed Nuclear Renaissance:

a Research Paper', Friends of the Earth, London, May. Available at: www.foe.co.uk/resource/reports/voodoo_economics.pdf.

Bruce Power (2009). 'Bruce Power to Focus on Additional Refurbishments at Bruce A and B', Press Release, 23 July. Available at: www.brucepower.com/pagecontent.aspx?navuid=1212&dtuid=84013.

Bunn, George (2007). 'Enforcing International Standards: Protecting Nuclear Materials From Terrorists Post-9/11', *Arms Control Today* (January/February). Available at: www.armscontrol.org/act/2007_01–02/Bunn.

Bunn, Matthew (2009). 'Reducing the Greatest Risks of Nuclear Theft & Terrorism', *Daedalus*, Vol. 138, No. 4, pp. 112–123.

Bunn, Matthew, Bob van der Zwaan, John P. Holdren and Steve Fetter (2003). 'The Economics of Reprocessing vs. Direct Disposal of Spent Nuclear Fuel', Project on Managing the Atom, Belfer Center for Science and International Affairs, John F. Kennedy School of Government, December.

Cadham, John (2009). 'The Canadian Nuclear Industry: Status and Prospects', *Nuclear Energy Futures Paper*, No. 8, 9 November. Available at: www.cigionline.org/publications/2009/11/canadian-nuclear-industry-status-and-prospects.

Canberra Commission (1996). 'Report of the Canberra Commission on the Elimination of Nuclear Weapons', Canberra: Department of Foreign Affairs. Available at: www.dfat.gov.au/cc/CCREPORT.PDF.

Central Intelligence Agency (2009). *The World Factbook.* Available at: www.cia.gov/library/publications/the-world-factbook.htm.

Charpin, Jean-Michel, Benjamin Dessus and René Pellat (2000). 'Economic Forecast Study of the Nuclear Power Option', Report to the Prime Minister, Paris, July. Available at: www.google.ca/search?q=Jean-Michel+Charpin%2C+Benjamin+Dessus+and+Ren%C3%A9+Pellat.

Choubey, Deepti (2008). 'Are New Nuclear Bargains Attainable?', Carnegie Endowment for International Peace, Washington, DC.

CNS (2009). 'Nuclear Suppliers Group', *Inventory of International Nonproliferation Organizations and Regimes*, James Martin Centre for Nonproliferation Studies, 15 June. Available at: www.nti.org/e_research/official_docs/inventory/pdfs/nsg.pdf.

CNSC (2008). *Annual Report 2007–08, Canadian Nuclear Safety Commission.* Ottawa: Ministry of Public Works and Services. Available at: www.nuclearsafety.gc.ca/pubs_catalogue/uploads/ar_2007_2008_e.pdf.

Cochran, Thomas B. (2007). 'Adequacy of IAEA's Safeguards for Achieving Timely Detection', in Henry Sokolski (ed.), *Falling Behind: International Scrutiny of the Peaceful Atom.* Washington, DC: Nonproliferation Policy Education Center.

Cochran, Thomas B. and Christopher E. Paine (2009). 'Proliferation Resistant Uranium Enrichment – Discussion Draft', presented at the International Forum on Nuclear Proliferation and Nuclear Energy, Amman.

Cohen, Roger (2008). 'America Needs France's Atomic Anne', *New York Times*, 24 January. Available at: www.nytimes.com.

Committee on International Security and Arms Control (CISAC) (1994). *Management and Disposition of Excess Weapons Plutonium.* Washington, DC: US National Academy of Sciences.

Commonwealth of Australia (2006). *Uranium Mining, Processing and Nuclear Energy-Opportunities for Australia?* ('The Switkowski Report'). Canberra: Department of the Prime Minister and Cabinet.

Congressional Budget Office (CBO) (2003). S.14: Energy Policy Act of 2003. Washington, DC: Congressional Budget Office Cost Estimate, 7 May. Available at: www.cbo.gov/doc.cfm?index=4206&type=0.

Congressional Budget Office (CBO) (2008). *Nuclear Power's Role in Generating Electricity*, CBO Report. Washington, DC: Congress of the United States. May.

Cooper, Mark (2009a). 'All Risk, No Reward for Taxpayers and Ratepayers: The Economics of Subsidizing the "Nuclear Renaissance" with Loan Guarantees and Construction Work in Progress', Vermont Law School, November. Available at: www.vermontlaw.edu/x9198.xml.

Cooper, Mark (2009b). 'The Economics of Nuclear Reactors: Renaissance or Relapse?', Institute for Energy and the Environment, Vermont Law School, June. Available at: www.nirs.org/neconomics/cooperreport_neconomics062009.pdf.

Cooper, Mark (2010). 'Cost Escalation, Diseconomics of Scale and Negative Learning in Nuclear Reactor Construction', Institute for Energy and the Environment, Vermont Law School, September.

Cravens, Gwyneth (2007). *Power to Save the World: the Truth About Nuclear Energy*. New York: Alfred A. Knopf.

Crawford, Ronald D. (2009). *Deputy Director, Regional Centre-Atlanta, World Association of Nuclear Operators at a Conference on Nuclear Power in Society: Finding the Balance*, Ottawa, 26 October.

Davis, Ian (2009). 'The British Nuclear Industry: Status and Prospects', *Nuclear Energy Futures Paper*, No. 4, January. Available at: www.cigionline.org.

Davis, Mary Byrd (2001). 'La France nucléaire: matières et sites 20002', Paris: WISE-Paris.

Dow Jones Newswires (2009). 'Areva, India Nuclear Power Corp in Pact for Reactors', *Wall Street Journal*, 9 February. Available at: http://online.wsj.com/article/SB123373837719447385.html.

Dowdeswell, Elizabeth (2005). Cover Letter, *Choosing a Way Forward: the Future Management of Canada's Used Nuclear Fuel: Final Study*. Toronto: Nuclear Waste Management Organization. November.

Dussart Desart, R. (2005). 'The Reform of the Paris Convention on Third Party Liability in the Field of Nuclear Energy and of the Brussels Supplementary Convention', *Nuclear Law Bulletin*, No. 75, 24.

Economist, The (2007). 'Egypt's Nuclear-Power Programme', 2 November. Available at: www.economist.com/displayStory.cfm?story_id=10085431.

Economist, The (2009a). 'Briefing on Nigeria, Hints of a New Chapter', 14 November, pp. 30–32.

Economist, The (2009b). 'Briefing: Smart Grids', 10 October.

Economist, The (2009c). 'On Target, Finally', 30 May.

Economist, The (2009d). 11 July, p. 45.

Economist, The (2010a). 'Atomic Dawn: KEPCO Wins a Nuclear Contract', 2 January, p. 47.

Economist, The (2010b). 'Electricity and Development in China: Lights and Action', 1 May.

Economist Technology Quarterly, The (2009). 6 June.

Egypt News (2008). 'Egypt to Assess First Nuclear Plant Tender Papers', 5 May. Available at: http://news.egypt.com/en/200805052457/news/-egypt-news/egypt-to-assess-first-nuclear-plant-tender-papers.html.

ElBaradei, Mohamed (2003a). Foreword, 'Periodic Safety Review of Nuclear Power Plants', IAEA Safety Guide, No. NS-G-2.10, IAEA, August.

ElBaradei, Mohamed (2003b). 'Statement to the Forty-seventh Regular Session of the IAEA General Conference 2003', 15 September. Available at: www.iaea.org/NewsCenter/Statements/2003/ebsp2003n020.html.

ElBaradei, Mohamed (2007). Foreword, 'The 1997 Vienna Convention on Civil Liability for Nuclear Damage and the 1997 Convention on Supplementary Compensation for Nuclear Damage: Explanatory Texts', IAEA International Law Series No. 3, IAEA document STI/PUB/1279, Vienna.

ElBaradei, Mohamed (2009). Statement to the Sixty-Fourth Regular Session of the United Nations General Assembly by IAEA Director General Dr. Mohamed ElBaradei, 2 November. Available at: www.iaea.org/NewsCenter/Statements/2009/ebsp2009n017.html.

Electric Power Research Institute (EPRI) (2002). *Deterring Terrorism: Aircraft Crash Impact Analyses Demonstrate Nuclear Power Plant's Structural Strength*, December. Available at: www.nei.org/filefolder/EPRI_Nuclear_Plant_Structural_Study_2002.pdf.

Elston, Murray (2009). 'Opening Remarks', speech given at the Canadian Nuclear Association Conference and Trade Show 2009 – The Reality of Renaissance, 26 February. Ottawa: Canadian Nuclear Association.

Energy Information Administration (2008). *International Energy Outlook 2008*. Washington, DC: Department of Energy, September. Available at: www.eia.doe.gov/oiaf/ieo/index.html.

Energy Information Administration (2009). *International Energy Outlook 2009*. Washington, DC: Department of Energy, September. Available at: www.eia.doe.gov/oiaf/ieo/index.html.

EurActiv.com (2010). 'Sweden Eyes Nuclear Revival after 30-year Ban', 23 March. Available at: www.euractiv.com/en/energy/sweden-eyes-nuclear-revival-after-30-year-ban-news-369566.

Faddis, Charles (2010). 'Al Qaeda's Nuclear Plant', *New York Times*, 6 May. Available at: www.nytimes.com/2010/05/06/opinion/06Faddis.html.

Feiveson, H.A. (2004). 'Nuclear Proliferation and Diversion', in Cutler Cleveland (ed.) *The Encyclopedia of Energy*. Elsevier.

Feiveson, Harold, Alexander Glaser, Marvin Miller and Lawrence Scheinman (2008). 'Can Future Nuclear Power Be Made Proliferation Resistant?', Center for International and Security Studies, University of Maryland, July. Available at: www.cissm.umd.edu/papers/files/future_nuclear_power.pdf.

Ferguson, Charles D. (2008). 'Strengthening Nuclear Safeguards', *Issues in Science and Technology*, Vol. 24, No. 3 (Spring). Available at: www.issues.org/24.3/ferguson.html.

Ferguson, Charles D. and Philip D. Reed (2009). 'Seven Principles of Highly Effective Nuclear Energy', in S. Apikyan and D.J. Diamong (eds) *Nuclear Power and Energy Security, NATO Science for Peace and Security Series – B: Physics and Biophysics*. Dordrecht: Springer, pp. 53–64.

Fischer, David (1997). *History of the International Atomic Energy Agency: The First Forty Years, International Atomic Energy Agency*. Vienna.

Fissile Material Working Group (FMWG) (2010). 'Next Generation Nuclear Security: Meeting the Global Challenge', Conference Report, May. Available at: www.stanley-foundation.org/resources.cfm?id=420.

Fox, Jon (2007). 'Standardize Nuclear Site Security, Auditors Say', *Global Security Newswire*, Nuclear Threat Initiative, 26 September. Available at: www.nti.org/d_newswire/issues/2007_9_26.html.

Friedrich, Andreas and Jim Finucane (2001) 'Guidelines for the Management of Plutonium (INFCIRC/549): Overview, Goals and Status', Addressing Excess Stocks of Civil and Military Plutonium, Institute for Science and International Security, 11 December. Available at: www.isis-online.org/uploads/conferences/documents/chap3.pdf.

Froggatt, Antony (2009). 'Dilute and Disperse', *NEI Magazine*, May.

G8 (2008). 'International Initiative on 3-S Based Nuclear Energy Infrastructure', G8 Summit Report, July. Available at: www.mofa.go.jp/policy/economy/summit/2008/doc/pdf/0708_04_en.pdf.

G8 (2009). 'G8 L'Aquila Statement on Non-Proliferation', 9 July. Available at: www.g7.utoronto.ca/summit/2009laquila/2009-nonproliferation.pdf.

Galbraith, Kate (2009). 'U.S.-Private Bid to Trap Carbon Emissions is Revived', *New York Times*, 15 June.

GAO (1979). 'Difficulties in Determining if Nuclear Training of Foreigners Contributes to Weapons Proliferation', Report by the Comptroller General of The United States, United States General Accounting Office.

GAO (2005). 'Nuclear Proliferation: IAEA has Strengthened its Safeguards and Nuclear Security Programs, but Weaknesses Need to be Addressed', GAO-06–93, Washington, DC, October.

GAO (2007). 'DOE and NRC Have Different Security Requirements for Protecting Weapons-Grade Material from Terrorist Attacks', US GAO document GAO-07–1197R, 11 September. Available at: www.gao.gov/new.items/d071197r.pdf.

GAO (2008). 'NRC's Oversight of Fire Protection at U.S. Commercial Nuclear Reactor Units Could Be Strengthened', US Government Accountability Office, Washington, DC, June. Available at: www.gao.gov/new.items/d08747.pdf.

Garwin, Richard L. and Georges Charpak (2001). *Megawatts + Megatons: The Future of Nuclear Power and Nuclear Weapons*. Chicago: University of Chicago Press.

GIF (2008). 'Generation IV Systems', Gen-IV International Forum. Available at: www.gen-4.org/Technology/systems/index.htm.

Gilinsky, Victor, Marvin Miller and Harmon Hubbard (2004). *A Fresh Examination of the Proliferation Dangers of Light Water Reactors*, The Nonproliferation Policy Education Center, 22 October. Available at: www.npecweb.org/Essays/20041022-GilinskyEtAl-LWR.pdf.

Global Security Newswire (2008a). 'Egypt to Accept Nuclear Construction Bids', *Global Security Newswire*, 28 January.

Global Security Newswire (2008b). 'Jordan Pursues Nuclear Trade Deals', *Global Security Newswire*, 30 July. Available at: www.nti.org/d_newswire/issues/2008_7_30.html#E0F15416.

Global Security Newswire (2008c). 'South Korea to Aid Jordanian Nuclear Program', *Global Security Newswire*, 1 December. Available at: http://gsn.nti.org/gsn/nw_20081201_2878.php.

Global Security Newswire (2010). 'India Submits Nuclear Liability Bill to Parliament', 7 May. Available at: www.globalsecuritynewswire.org/gsn/nw_20100507_6708.php.

Goldblat, Jozef (2002). *Arms Control: the New Guide to Negotiations and Agreements*. London: Sage.

Goldenberg, Suzanne (2010). 'Barack Obama Gives Green Light to New Wave of Nuclear Reactors', *Guardian*, 23 February.

Goldschmidt, Pierre (2007). 'Saving the NPT and the Nonproliferation Regime in an Era of Nuclear Renaissance', Testimony before the House Foreign Affairs Subcommittee on Terrorism, Nonproliferation and Trade, 28 July.

Goldschmidt, Pierre (2009). 'Concrete Steps to Improve the Nonproliferation Regime', *Carnegie Paper*, No. 100, April.

Gonzáles, Abel J. (2002). 'International Negotiations on Radiation and Nuclear Safety', in R. Avenhaus, V. Kremenyuk and G. Sjöstedt (eds) *Containing the Atom: International Negotiations on Nuclear Security and Safety*. Lanham: Lexington Books.

Gordon, Jeremy (2009). 'Nuclear Shutdown in Brazil Blackout', *World Nuclear News*, 12 November.

Gourley, Bernard and Adam N. Stulberg (2009). 'Nuclear Energy Development: Assessing Aspirant Countries', *Bulletin of the Atomic Scientists*, Vol. 65, No. 6, November/December, pp. 20–29.

Gregoric, Miroslav (2009). 'IAEA Activities in Nuclear Security', Department of Nuclear Safety and Security, Vienna, IAEA, 7 October. Available at: www.iaea.or.at/OurWork/ST/NE/NEFW/CEG/documents/CEG_23/1–9%20IAEA%20Eng.pdf.

Grossman, Elaine M. (2009). 'Boost in IAEA Intelligence Capability Looks Unlikely in Near Term', *Global Security Newswire*, 22 June.

Grunwald, Michael (2010). 'Why Obama's Nuclear Bet Won't Pay Off', *Time Magazine*, 18 February. Available at: www.time.com/time/politics/article/0,8599,1964846,00.html.

Gulf News (2008). 'Algeria Plans Law for Nuclear Power this Year', *GulfNews.com*, 1 July. Available at: www.gulfnews.com/region/Algeria/10225217.html.

Hamid, Asma (2008). 'N-power to Meet 15 pc of UAE Energy Needs', *Khaleej Times Online*, 25 November. Available at: www.khaleejtimes.ae/DisplayArticle.asp?xfile=data/theuae/2008/November/theuae_November 557.xml&section=theuae.

Handl, Gunther (2003). 'The IAEA Nuclear Safety Conventions: an Example of Successful 'Treaty Management'?', *Nuclear Law Bulletin*, No. 72.

Hannum, William H., Gerald E. Marsh and George S. Stanford (2005). 'Smarter Use of Nuclear Waste', *Scientific American*, December.

Hansen, James (2008). 'Target Atmospheric CO2: Where Should Humanity Aim?', *Open Atmospheric Science Journal*, Vol. 2, pp. 217–231.

Harisumarto, Sukino (2007). 'Villagers Against Indonesia's Plans for Nuclear Power Plant', *Nuclear Features*, 26 August. Available at: www.monstersandcritics.com/news/energywatch/nuclear/features/article_1347879.php/Villagers_against_Indonesias_plans_for_nuclear_power_plant.

Herbst, Alan M. and George W. Hopley (2007). *Nuclear Energy Now: Why The Time Has Come for the World's Most Misunderstood Energy Source*. Hoboken: John Wiley & Sons Inc.

Heupel, Monika (2007). 'Implementing UN Security Council Resolution 1540: A Division of Labor Strategy', Carnegie Endowment for International Peace, Nonproliferation Program, No. 87, Washington, DC, June. Available at: www.carnegieendowment.org/files/cp87_heupel_final.pdf.

Hibbs, Mark (2008). 'Potential Personnel Shortages Loom Over China's Nuclear Expansion', *Nucleonics Week*, Vol. 49, No. 24.

Hibbs, Mark (2009). 'Bahrain Moving Cautiously on Nuclear Energy Deliberations', *Nucleonics Week*, 12 March.

Hiruo, Elaine (2009). 'Sandia Seeks Industry Partner for Small Reactor', *Nucleonics Week*, 3 September.

Horner, David (2008). 'NFS Executive Sees Progress, Obstacles in Safety Culture Effort', *Inside NRC*, 23 June.

House of Lords Science and Technology Committee (2007). 'Radioactive Waste Management: an Update', The Stationery Office, House of Lords Science and Technology Committee, 4th Report of 2006/07, HL Paper 109.

Howsley, Roger (2009). 'The World Institute for Nuclear Security: Filling a Gap in the Global Nuclear Security Regime', *Innovations*, Vol. 4, No. 4 (Fall 2009), pp. 203–208.

Huntley, Wade and Karthika Sasikumar (eds) (2006). *Nuclear Cooperation with India: New Challenges, New Opportunities, Simons Centre for Disarmament and Non-Proliferation Research*. University of British Columbia, Vancouver.

Hutton, John (2008). Statement by Business and Enterprise Minister John Hutton, quoted in 'Nuclear Power is "Vital" to Britain', *World Nuclear News*, 11 January.

IAEA (1972). 'The Structure and Content of Agreements Between the Agency and States Required in Connection with the Treaty on the Non-Proliferation of Nuclear Weapons (NPT)', IAEA document INFCIRC/153 (Corrected), June.

IAEA (1973). 'Market Survey for Nuclear Power in Developing Countries: General Report', Vienna, September.

IAEA (1974). 'The Standard Text of Safeguards Agreements in Connection with the Treaty on the Non-Proliferation of Nuclear Weapons', IAEA document GOV/INF/276, 22 August.

IAEA (1993). 'Safety Fundamentals: the Safety of Nuclear Installations', Safety Series No. 110.

IAEA (1994). 'Convention on Nuclear Safety: Final Act', IAEA document INFCIRC/449/Add.1, 4 August.

IAEA (1997). 'Provision of Safety Related Assistance Through the Agency's Technical Co-Operation Programme' (Part C, Annex C-1), 'Measures to Strengthen International Co-operation in Nuclear, Radiation and Waste Safety', IAEA document GC(41)/INF/8, 2 September.

IAEA (1998). 'The Physical Protection of Nuclear Material and Nuclear Facilities', IAEA document INFCIRC/225/Rev. 4 (Corrected).

IAEA (1999). 'Guidelines Regarding National Reports Under the Convention on Nuclear Safety', IAEA document INFCIRC/572/Rev. 1, 15 October.

IAEA (2003a). *Handbook on Nuclear Law*, IAEA publication STI/PUB/1160, July.

IAEA (2003b). 'Periodic Safety Review of Nuclear Power Plants', Safety Guide, No. NS-G-2.10, IAEA, August.

IAEA (2003c). *PROSPER guidelines: Guidelines for Peer Review and for Plant Self-assessment of Operational Experience Feedback Process*, April.

IAEA (2003d). Report to the Board of Governors by the Co-Chairmen of the Informal Open-ended Working Group on the Programme and Budget for 2004–2005, IAEA document GC(47)INF/7, Attachment 1, 16 July.

IAEA (2004). 'Guidelines for the Management of Plutonium (INFCIRC/549): Background and Declarations', 1 April (revised 16 August 2005). Available at: www.isis-online.org/uploads/conferences/documents/chap3.pdf.

IAEA (2005). 'OSART: Operational Safety Review Teams', IAEA. Available at: www-ns.iaea.org/downloads/ni/s-reviews/osart/OSART_Brochure.pdf.

IAEA (2006a). 'ITDB Fact Sheet', Vienna. Available at: www.iaea.org/NewsCenter/Features/RadSources/PDF/fact_figures2006.pdf.

IAEA (2006b). *Main Elements of the Global Nuclear Safety Regime, Strengthening the Global Nuclear Safety Regime*, INSAG Report 21 (INSAG-21).

IAEA (2006c). 'Measures to Strengthen International Cooperation in Nuclear, Radiation and Transport Safety and Waste Management', IAEA resolution GC(50)/Res/1, September.

IAEA (2006d). 'Nuclear Security – Measures to Protect Against Nuclear Terrorism',

Annual Report, Director General, IAEA document GOV/2006/46-GC(50)/13, 16 August.

IAEA (2006e). 'Revision of the Standardized Text of the "Small Quantities Protocol"', IAEA document GOV/INF/276/Mod.1, 21 February.

IAEA (2007a). 'Considerations to Launch a Nuclear Power Programme', IAEA document GOV/INF/2007/2, April.

IAEA (2007b). 'Energy, Electricity and Nuclear Power: Developments and Projections – 25 Years Past & Future', Vienna.

IAEA (2007c). 'Milestones in the Development of a National Infrastructure for Nuclear Power', IAEA Nuclear Energy Series, no. NG-G-3.1.

IAEA (2007d). 'Safeguards Analytical Laboratory: Sustaining Credible Safeguards, Report by the Director General', IAEA document GOV/2007/59, 24 October.

IAEA (2007e). 'The Vienna Convention on Civil Liability for Nuclear Damage and the 1997 Convention on Supplementary Compensation for Nuclear Damage – Explanatory Texts', IAEA, International Law Series, No. 3, IAEA document STI/PUB/1279, Vienna.

IAEA (2008a). '20/20 Vision for the Future: Background Report by the Director General for the Commission of Eminent Persons', IAEA document GOV/2008-GC(52)/INF/4, May.

IAEA (2008b). 'An Agreement with the Government of India for the Application of Safeguards to Civilian Nuclear Facilities', *Nuclear Verification: The Conclusion of Safeguards Agreements and Additional Protocols*, IAEA document GOV/2008/30, 9 July. Available at: www.isis-online.org/publications/southasia/India_IAEA_safeguards. pdf.

IAEA (2008c). 'Annual Report 2007', IAEA document GC(52)/9, Vienna.

IAEA (2008d). 'Classification of Radioactive Waste' (Draft DS-390), IAEA Safety Standards Series, 27 November.

IAEA (2008e). *Energy, Electricity, and Nuclear Power Estimates for the Period up to 2030*, Vienna, September.

IAEA (2008f). 'Evaluation of the Status of National Nuclear Infrastructure Development', IAEA Nuclear Energy Series, no. NG-T-3.2.

IAEA (2008g). 'International Nuclear Security Advisory Service', 30 June. Available at: www-ns.iaea.org/security/insserv.htm.

IAEA (2008h). 'International Physical Protection Advisory Service', 30 June. Available at: www-ns.iaea.org/security/ippas.htm.

IAEA (2008i). 'Nuclear Power Worldwide: Status and Outlook', Press Release, International Atomic Energy Agency, 11 September. Available at: www.iaea.org/NewsCenter/PressReleases/2008/prn200811.html.

IAEA (2008j). 'Nuclear security – Measures to Protect Against Nuclear Terrorism', Annual Report 2007, IAEA document GC(52)/9, March.

IAEA (2008k). 'Plan of Action to Promote the Conclusion of Safeguards Agreements and Additional Protocols', Vienna, September. Available at: www.iaea.org/OurWork/SV/Safeguards/sg_actionplan.pdf.

IAEA (2008l). *Reinforcing the Global Nuclear Order for Peace and Prosperity: The Role of the IAEA to 2020 and Beyond*, report prepared by an independent Commission at the request of the Director-General of the IAEA, IAEA document GOV/2008/22-GC(52)/INF/4, Annex 1, Vienna, May.

IAEA (2008m). *Summary Report of the 4th Review Meeting of the Contracting Parties to the Convention on Nuclear Safety*, IAEA document CNS/RM/2008–6 Final, April.

IAEA (2009a). 'Convention on Assistance in the Case of a Nuclear Accident or Radio-logical Emergency: Status', IAEA, 31 October. Available at: www.iaea.org/Publications/Documents/Conventions/cacnare_status.pdf (accessed 11 December 2009).

IAEA (2009b). 'Convention on Early Notification of a Nuclear Accident: Status', IAEA, 31 October. Available at: www.iaea.org/Publications/Documents/Conventions/cenna_status.pdf (accessed 11 December 2009).

IAEA (2009c). 'Convention on Nuclear Safety: Status', IAEA, 13 August. Available at: www.iaea.org/Publications/Documents/Conventions/nuclearsafety_status.pdf.

IAEA (2009d). 'Disposable Waste Management', IAEA. Available at: www-ns.iaea.org/tech-areas/waste-safety/disp-book.htm.

IAEA (2009e). 'Finding a Role for Nuclear: IAEA Helps Developing Countries Assess Readiness for Nuclear Power', Staff Report, 21 July. Available at: www.iaea.org/NewsCenter/News/2009/nuclearrole.html.

IAEA (2009f). 'IAEA by the Numbers', Vienna. Available at: www.iaea.org/About/by_the_numbers.html (accessed 15 December 2009).

IAEA (2009g). 'INPRO – International Project on Innovative Nuclear Reactors and Fuel Cycles'. Available at: www.iaea.org/INPRO/.

IAEA (2009h). 'Implementation of the IAEA Nuclear Security Plan 2006–2009: Progress Report', IAEA document GOV/2009/53-GC(53)16, September.

IAEA (2009i). 'International Conventions and Agreements'. Available at: www.iaea.org/Publications/Documents/Conventions/index.html.

IAEA (2009j). 'Nuclear Safety Review for the Year 2008', IAEA document GD(53)/INF/2, July.

IAEA (2009k). 'Nuclear Security, Including Measures to Protect Against Nuclear and Radiological Terrorism', IAEA document GC(53)/RES/11, September.

IAEA (2009l). 'OSART Mission List'. Available at: www-ns.iaea.org/downloads/ni/s-reviews/osart/osart%20mission%20list%202009.pdf.

IAEA (2009m). 'President's Findings from the International Symposium on Nuclear Security', 3 April. Available at: www-pub.iaea.org/mtcd/meetings/PDFplus/2009/cn166/cn166_SymposiumFinding.doc.

IAEA (2009n). 'Regular Budget Appropriations for 2010', IAEA document GC(53)/RES/6, 18 September.

IAEA (2009o). 'Safety Standards Commission and Committees', IAEA. Available at: www-ns.iaea.org/committees/.

IAEA (2009p). 'States Briefed on Sustainable Nuclear Future', 18 September. Available at: www.iaea.org/NewsCenter/News/2009/inpro.html.

IAEA (2009q). 'States' Participation in Major Agreements, Office of Legal Affairs', 22 December. Available at: http://ola.iaea.org/lars/ReportOutput/GlobalReport.pdf.

IAEA (2009r). 'Status and Near Term Prospects of Small and Medium Sized Reactors', November. Available at: www.iaea.org/NuclearPower/Downloads/SMR/docs/Status-SMRs-January-2010.pdf.

IAEA (2009s). *Summary Report of the 3rd Review Meeting of the Contracting Parties to the Joint Convention on the Safety of Spent Fuel and on the Safety of Radioactive Waste Management*, IAEA document JC/RM3/02/Rev2, 20 May.

IAEA (2009t). 'TC Programme: Projects by Field', International Atomic Energy Agency. Available at: www-tc.iaea.org/tcweb/tcprogramme/projectsbyfacandapc/default.

IAEA (2009u). 'The Agency's Programme and Budget 2010–2011', GC(53)/5, International Atomic Energy Agency, August.

IAEA (2009v). 'Nuclear Research Reactors in the World', December. Available at: www. iaea.org/worldatom/rrdb/.

IAEA (2010a). 'Agreement Reached on Integrated Safeguards in European Union', Press Release 2010/01. Available at: www.iaea.org/NewsCenter/PressReleases/2010/prn201001.html.

IAEA (2010b). 'Conclusion of Safeguards Agreements, Additional Protocols and Small Quantities Protocols', 27 May. Available at: www.iaea.org/OurWork/SV/Safeguards/sir_table.pdf.

IAEA (2010c). *Power Reactor Information Service (PRIS)*. Available at: www.iaea.org/programmes/a2/.

IAEA (2010d). 'States' Participation in Major Agreements, Office of Legal Affairs', 1 February. Available at: http://ola.iaea.org/lars/ReportOutput/GlobalReport.pdf.

IAEA (2010e). 'Status of the Joint Convention on the Safety of Spent Fuel Management and on the Safety of Radioactive Waste Management', 29 April. Available at: www. iaea.org/Publications/Documents/Conventions/jointconv_status.pdf.

ICNND (2009). *Eliminating Nuclear Threats: A Practical Agenda for Global Policymakers*, International Commission on Nuclear Non-Proliferation and Disarmament. Canberra, November.

INPO (2009). 'Institute of Nuclear Power Operations (INPO)'. Available at: http://en.wikipedia.org/wiki/Institute_of_Nuclear_Power_Operations.

Inside NRC (2008). 'France to Establish Criteria for New Cooperative Agreements', 23 June.

Intergovernmental Panel on Climate Change (IPCC) (2007). *Climate Change 2007: Synthesis Report*, Geneva. Available at: www.ipcc.ch/pdf/assessment-report/ar4/syr/ar4_syr.pdf.

International Energy Agency (IEA) (2008a). *Energy Technology Perspectives 2008: Scenarios and Strategies to 2050*. Paris: Organisation for Economic Cooperation and Development (OECD).

International Energy Agency (IEA) (2008b). *World Energy Outlook 2008*. Paris: OECD/IEA.

International Energy Agency (IEA) (2009a). 'The Impact of the Financial and Economic Crisis on Global Energy Investment', International Energy Agency Background Paper for the G8 Energy Ministers Meeting, Executive Summary. Rome, 24–25 May. Available at: www.iea.org/Papers/2009/G8_investment_ExecSum.pdf.

International Energy Agency (IEA) (2009b). *World Energy Outlook 2009*. Paris: OECD/IEA.

International Institute for Strategic Studies (IISS) (2008). *Nuclear Programmes in the Middle East: In the Shadow of Iran*. London: IISS Strategic Dossier.

International Institute for Strategic Studies (IISS) (2009). 'Preventing Nuclear Dangers in Southeast Asia and Australasia', Press statement. London: International Institute for Strategic Studies.

International Safety Advisory Group (INSAG) (2006a). 'Foreword by the Chairman of INSAG', *Strengthening the Global Nuclear Safety Regime*, INSAG-21, IAEA.

International Safety Advisory Group (INSAG) (2006b). *Strengthening the Global Nuclear Safety Regime*, INSAG-21, IAEA.

IPFM (2007). 'Global Fissile Material Report 2007', October. Available at: www.fissile-materials.org/ipfm/site_down/gfmr07.pdf, 2007.

IPFM (2009). 'A Path to Nuclear Disarmament', *Global Fissile Material Report 2009, International Panel on Fissile Materials*, October. Available at: www.fissilematerials.org/ipfm/site_down/gfmr09.pdf.

Jakarta Post (2007). 'Batan Seeks Decree on Indonesian Plant', 6 December. Available at: www.world-nuclear-news.org/industry/Batan_seeks_decree_on_Indonesian_plant-061207.shtml.

Joskow, Paul L. and John E. Parsons (2009). 'The Economic Future of Nuclear Power', *Daedalus*, Vol. 138, No. 4, pp. 45–59.

KEPCO (2009). KEPCO 2009 Annual Report. Available at: http://multi.kepco.co.kr/annual/2009/kepco_eng.pdf.

Kerr, Paul (2004). 'IAEA Probes Seoul's Nuclear Program', *Arms Control Today*, October. Available at: www.armscontrol.org/act/2004_10/IAEA_Seoul_Nuclear_Program.

Kerr, Paul (2007). 'ElBaradei: IAEA Budget Problems Dangerous', *Arms Control Today*, July/August. Available at: www.armscontrol.org/act/2007_07–08/IAEABudget.

Kessler, Carol and Lindsay Windsor (2007). *Technical and Political Assessment of Peaceful Nuclear Power Program Prospects in North Africa and the Middle East*. Richland, Washington: Pacific Northwest National Laboratory.

Keystone Center (2007). *Nuclear Power Joint Fact-Finding Final Report*, Keystone, June. Available at: http://keystone.org/files/file/about/publications/FinalReport_NuclearFactFinding6_2007.pdf.

Khanh, Vu Trong and Patrick Barta (2009). 'Vietnam Assembly Approves Nuclear Plants', *Wall Street Journal*, 26 November. Available at: http://online.wsj.com/article/SB125913876138763683.html.

Kidd, Steve (2008). *Core Issues: Dissecting Nuclear Power Today*. Sidcup: Nuclear Engineering International Special Publications.

Kidd, Steve (2009a). 'New Nuclear Build – Sufficient Supply Capability?', *Nuclear Engineering International*, 3 March. Available at: www.neimagazine.com/story.asp?storyCode=2052302.

Kidd, Steve (2009b). 'New Nuclear Power Plants: an Insurmountable Risk?', *NEI Magazine*, 13 April.

Klein, Dale E. (2007). 'Remarks As Prepared for NRC Chairman Dale E. Klein' (No. 07–041), U.S. Nuclear Regulatory Commission, 10 September. Available at: http://adamswebsearch2.nrc.gov/idmws/doccontent.dll?library=PU_ADAMS^PBNTAD01&ID=072530545.

Kramer, Andrew E. (2010). 'Safety Issues Linger as Nuclear Reactors Shrink in Size', *New York Times*, 18 March. Available at: www.nytimes.com/2010/03/19/business/energy-environment/19minireactor.html.

Kroenig, Matthew (2009). 'Exporting the Bomb: Why States Provide Sensitive Nuclear Assistance', *American Political Science Review*, Vol. 103, No. 1.

Kumar, Ashwin and M.V. Ramana (2008). 'Compromising Safety: Design Choices and Severe Accident Possibilities in India's Prototype Fast Breeder Reactor', *Science and Global Security*, Vol. 16, Issue 3, pp. 87–114.

Kumar, Himendra Mohan (2008). 'Nuclear Matters', *Gulf News*, 31 July. Available at: www.gulfnews.com/business/General/10233061.html.

Lochbaum, David (2006). 'Walking a Nuclear Tightrope: Unlearned Lessons of Year-plus Reactor Outages', *Union of Concerned Scientists*, September.

Loukianova, Anya (2008). 'Issue Brief: The International Uranium Enrichment Center at Angarsk: A Step Towards Assured Fuel Supply?', *Nuclear Threat Initiative*, November. Available at: www.nti.org/e_research/e3_93.html.

Lovins, Amory B. and Imran Sheikh (2008). 'The Nuclear Illusion', *Draft Manuscript for Ambio*, November edition.

Lowbeer-Lewis, Nathaniel (2010) 'Nigeria and Nuclear Energy: Plans and Prospects', *Nuclear Energy Futures Paper*, No. 10. Available at: www.cigionline.org.

McBride, J.P., R.E. Moore, J.P. Witherspoon and R.E. Blanco (1978). 'Radiological Impact of Airborne Effluents of Coal and Nuclear Plants', *Science*, Vol. 202, No. 4372, pp. 1045–1050.

McDonald, Allan and Leo Schrattenholzer (2001). 'Learning Rates for Energy Technologies', *Energy Policy*, No. 29, pp. 355–261.

MacKay, David J.C. (2009). *Sustainable Energy Without the Hot Air.* Cambridge: UIT.

MacKay, Paul (1998). 'Why Candus are Bomb Kits', *The Ottawa Citizen*, 7 June. Available at: http://energy.probeinternational.org/nuclear-power/nuclear-proliferation/why-candus-are-bomb-kits.

McKeeby, David (2008). 'White House Withdraws Russian Nuclear Agreement from Congress: Multibillion-dollar Deal on Hold Over Georgia Crisis, says White House', America.gov, 8 September. Available at: www.america.gov/st/peacesec-english/2008/September/20080908175454idybeekcm0.6636316.html.

MacLachlan, Ann (2007a). 'Better International Reporting Surfaces as a New Push for INSAG', *Inside NRC*, 1 October, pp. 10–11.

MacLachlan, Ann (2007b). 'IAEA Workshop to Pursue "Holistic" Approach to Radwaste Management', *NuclearFuel*, 2 July, p. 10.

MacLachlan, Ann (2008a). 'ASN Tells Areva it Needs Review of Corporate Safety Management', *NuclearFuel*, 25 August, pp. 3–4.

MacLachlan, Ann (2008b). 'French Agency, Four Arab Countries Discuss Nuclear Cooperation', *Nucleonics Week*, 25 September.

MacLachlan, Ann (2008c). 'Newcomers to Nuclear Power Urged to Join Nuclear Safety Convention', *Nucleonics Week*, 17 April, p. 10.

MacLachlan, Ann (2008d). 'Regulatory Independence Key Topic as Safety Convention Parties Meet', *Nucleonics Week*, 1 May, p. 17.

MacLachlan, Ann (2008e). 'US Ratification Boosts Plan for International Nuclear Liability', *Nucleonics Week*, 3 January, p. 6.

MacLachlan, Ann (2009a). 'Areva's MH1's Atmea1 Midsize PWR said Ready for Bidding in Jordan', *Nucleonics Week*, 19 March.

MacLachlan, Ann (2009b). 'Financial Crisis Nips Nuclear Revival in the Bud, WNA Told', *Nucleonics Week*, 17 September.

MacLachlan, Ann (2009c). 'GEH: Cost Estimates did Industry a "Disservice"', *Nucleonics Week*, 17 September, p. 1.

MacLachlan, Ann (2009d). 'Klein: Reactor Life beyond 80 is Technological, Human Challenge', *Inside NRC*, 14 September, p. 3.

MacLachlan, Ann (2009e). 'Multinational Design Certification Seen as Easier than Once Thought', *Inside NRC*, 26 September, p. 6.

MacLachlan, Ann and Mark Hibbs (2009). 'Westinghouse Seeks Chinese Consent to Design Changes on AP1000', *Nucleonics Week*, 2 April, pp. 1, 9–11.

McLellan, David (2007). 'The Economics of Nuclear Power', *Nuclear Energy Futures Paper*, No. 1, September. Available at: www.cigionline.org/sites/default/files/Economics%20of%20Nuclear%20Power.pdf.

Marshall, Pearl (2007). 'Most Nuclear Countries Undecided on Back-end Options, WEC Says', *NuclearFuel*, 24 September, p. 17.

Massachusetts Institute of Technology (MIT) (2003). *Future of Nuclear Power*. Boston, MIT.

Massachusetts Institute of Technology (MIT) (2009). *Update of the MIT 2003 Future of Nuclear Power*, Boston, MIT.

May, Kathryn (2010). 'Layoff of AECL Employees Badly Timed, Union Says', *Ottawa Citizen*, 1 June.

Mee-young, Cho (2010). 'Factbox – S. Korea's Nuclear Power Reactor Profiles', Reuters. Available at: www.reuters.com/article/idUKTOE64602320100513.

Merabet, Y. (2009). 'Le nucléaire Algérien et le bavardage de Chakib Khelil (1ère partie)', *Le Matin*, 4 August. Available at: www.lematindz.net/news/2735-le-nucleaire-algerien-et-le-bavardage-de-chakib-khelil-1ere-partie.html.

Meserve, Richard A. (2009) 'The Global Nuclear Safety Regime', *Daedalus*, Vol. 138, No. 4, pp. 100–111.

Miller, Marvin (2004). 'The Feasibility of Clandestine Reprocessing of LWR Spent Fuel', Appendix 2, in Victor Gilinsky *et al.*, *A Fresh Examination of the Proliferation Dangers of Light Water Reactors*.

Moj News Agency (2008). 'Algeria Supports Iran's Nuclear Plan', 1 July. Available at: www.mojnews.com/en/news_full_story.asp?nid=3243.

Moody's Corporate Finance (2007). 'New Nuclear Generation in the United States', *Special Comment*, October.

Moore, Patrick (2006). 'Going Nuclear', *The Washington Post*, 16 April.

Mozley, Robert F. (1998). *The Politics and Technology of Nuclear Proliferation*. Washington, DC: University of Washington Press.

Mueller, John (2010). *Atomic Obsession: Nuclear Alarmism from Hiroshima to Al-Qaeda*. Oxford: Oxford University Press.

Muroya, Nobuiho (2009). Director, Division of Operations C, Department of Safeguards, IAEA, to Wilton Park Conference 1008 on 'Nuclear Non-Proliferation and the 2010 Review', 14–18 December.

National Academy of Sciences (NAS) (2009). *Internationalization of the Fuel Cycle: Goals, Strategies and Challenges*. Washington, DC: The National Academies Press.

New York Times (2010). 'New York Denies Indian Point a Permit', 4 April. Available at: www.nytimes.com/2010/04/04/nyregion/04indian.html?fta=y.

NNSA (2009). 'Next Generation Safeguards Initiative Annual Report for FY2009', Office of Nonproliferation and Security, National Nuclear Security Administration, Department of Energy, Washington, DC.

Norris, Robert C. (2000). *The Environmental Case for Nuclear Power: Economic, Medical, and Political Considerations*. St. Paul: Paragon House.

NRC (2009a). 'Backgrounder on the Three Mile Island Accident', 13 March. Available at: www.nrc.gov/reading-rm/doc-collections/fact-sheets/3mileisle.html.

NRC (2009b). 'Safety Evaluation Report: Related to the License Renewal of Indian Point Nuclear Generating Unit Nos. and 2, 3' (Dockets 50-and 247, 50–286), Office of Nuclear Reactor Regulation, US Nuclear Regulatory Commission, August. Available at: http://adamswebsearch2.nrc.gov/idmws/ViewDocByAccession.asp?AccessionNumber=ML092240268.

NRC (2009c). 'NRC Issues Final Rule on New Reactor Impact Assessment', *News Release*. Washington, DC: US Nuclear Regulatory Commission. 17 February.

NRC News (2009). 'NRC Issues Final Rule on New Reactor Impact Assessment', *NRC News*, 17 February.

NRCan (2009). 'Government of Canada Invites Investor Proposals for AECL CANDU Reactor Division', Press Release No. 2009/123, Natural Resources Canada, 17 December. Available at: www.nrcan-rncan.gc.ca/media/newcom/2009/2009123–1a-eng.php.

Nuclear Energy Agency (NEA) (2000). 'Nuclear Education and Training: Cause for Concern? A Summary Report'. Paris: OECD.

Nuclear Energy Agency (NEA) (2004). 'Nuclear Competence Building: Summary Report', NEA No. 5588. Paris: NEA/OECD.

Nuclear Energy Agency (NEA) (2005). *Projected Costs of Generating Electricity*. Paris: OECD Publishing.

Nuclear Energy Agency (NEA) (2008). *Nuclear Energy Outlook 2008*, NEA No. 6348. Paris: OECD.

Nuclear Energy Agency (NEA) (2009a). *Multinational Design Evaluation Programme Annual Report*. Paris.

Nuclear Energy Agency (NEA) (2009b). Nuclear Energy Agency home page, OECD. Available at: www.nea.fr/.

Nuclear Energy Agency (NEA) (2009c). *Nuclear Energy Data*. Paris: OECD.

Nuclear Energy Daily (2007). 'Nuclear Power Share-Out not Delaying Grid Deal: Lithuania, Poland', 10 October. Available at: www.energy-daily.com.

Nuclear Energy Institute (NEI) (2009). 'The Cost of New Generating Capacity in Perspective', *NEI White Paper*, February. Washington, DC: Nuclear Energy Institute.

Nuclear Energy Policy Study Group (1977). *Nuclear Power Issues and Choices, Report of the Nuclear Energy Policy Study Group*. Cambridge: Ballinger Publishing Company.

Nuclear Engineering International (2007). 'Nuclear Waste: Without the Fanfare', July.

Nuclear Fuel (2008). 'US-ROK Negotiation Will be Key to Korea's Closed Fuel Cycle Plans', 5 May.

Nuclear Fuel Cycle (2008). 'Budget Shortfall Puts Yucca Site in "Standby"', 14 January.

Nuclear News Flashes (2007a). 'High Nuclear Plant Capital Costs May Discourage even Developed Countries from Building New Plants', *Platts*, 19 September.

Nuclear News Flashes (2007b). *Platts*, 11 September.

Nuclear News Flashes (2007c). *Platts*, 26 October.

Nuclear News Flashes (2009a). *Platts*, 17 March.

Nuclear News Flashes (2009b). *Platts*, 18 March.

Nuclear News Flashes (2009c). *Platts*, 24 March.

Nuclear News Flashes (2009d). *Platts*, 8 July.

Nuclear News Flashes (2009e). *Platts*, 23 September.

Nuclear Power Joint Fact-Finding Report (2007). Keystone, CO: Keystone Center, June. Available at: www.keystone.org/spp/energy/electricity/nuclear-power-dialogue (accessed 15 December 2009).

Nuclear Regulatory Commission, US Department of Energy (2009). 'Expected New Nuclear Power Plant Applications', 28 September. Available at: www.nrc.gov/reactors/new-reactors/new-licensing-files/expected-new-rx-applications.pdf.

Nuclear Threat Initiative (NTI) (2009a). 'NTI in Action: Creating an International Nuclear Fuel Bank', Nuclear Threat Initiative. Available at: www.nti.org/b_aboutnti/b7_fuel_bank.html.

Nuclear Threat Initiative (NTI) (2009b). 'North Korea Profile: Nuclear', *Country Profiles*, Nuclear Threat Initiative, December. Available at: www.nti.org/e_research/profiles/NK/index.html.

Nucleonics Week (2007a). 'No Ignalina-2 Replacement Near; Lithuania May Revisit Shutdown', *Platts*, 18 October.

Nucleonics Week (2007b). 'Politics and Calamities Stalking Jakarta's Nuclear Power Ambitions', *Platts*, 27 September.

Nucleonics Week (2008). 'EDF Expected to Announce 20% Rise in Projected Costs of Flamanville-3', *Platts*, 4 December.

Nucleonics Week (2009a). 'Olkiluoto-3 Manager Says Engineering Judgement Undermined', *Platts*, 26 March.

Nucleonics Week (2009b). 'India Seen Moving to Choose Sites for US Reactors Soon', *Platts*, 5 February.

Nucleonics Week (2009c). 'Swedish Opposition Reaffirms Gradual Nuclear Phase-out Policy', *Platts*, 26 March.

Nucleonics Week (2010). 'Russian Industry to Build Vietnam's First Nuclear Plant', *Platts*, 29 April.

NUKEM Inc. (2008). 'Nuclear Renaissance: USA Coping with New NPP Sticker Shock', *NUKEM Market Report*, April.

Nuttall, W.J. (2005). *Nuclear Renaissance: Technologies and Policies for the Future of Nuclear Power*. New York: Taylor & Francis Group.

NWMO (2005). *Choosing a Way Forward: The Future Management of Canada's Used Nuclear Fuel, Nuclear Waste Management Organization* (Canada), Toronto, November.

Oberth, Ron (2009). 'Written Submission by Atomic Energy of Canada Limited to Saskatchewan Crown & Central Agencies Energy Options Committee Meeting', *Atomic Energy of Canada Limited (AECL)*, 14 October. Available at: www.legassembly.sk.ca/committees/CrownCentralAgencies/Tabled%20Documents/AECL_submission.pdf.

Ontario Clean Air Alliance (2000). 'Air Quality Issues Fact Sheet', No. 26, 7 May. Available at: www.cleanairalliance.org/resource/fs26.pdf.

Oshima, Kenichi (2009). 'Nuclear Power in Japan – Top Priority of National Energy Policy', in L. Metz *et al.* (eds) *International Perspectives on Energy Policy and the Role of Nuclear Power*. Brentwood: Multi-Science Publishing.

OTA (1995). *Nuclear Safeguards and the International Atomic Energy Agency*. Office of Technology Assessment (OTA), OTA-ISS-615, US Congress, Washington, DC, April.

Pacala, S. and R. Socolow (2004). 'Stabilization Wedges: Solving the Climate Problem for the Next 50 Years with Current Technologies', *Science*, 13 August.

Paviet-Hartmann, Patricia, Bob Benedict and Michael J. Lineberry (2009). 'Nuclear Fuel Reprocessing', in Kenneth D. Kok (ed.) *Nuclear Engineering Handbook*. Boca Raton: CRC Press.

Perkovich, George (1999). *India's Nuclear Bomb: the Impact on Global Proliferation*. Berkeley: University of California Press.

Perkovich, George (2002). 'Nuclear Power in India, Pakistan and Iran', in Paul Leventhal, Sharon Tanzer and Steven Dolley (eds) *Nuclear Power and the Spread of Nuclear Weapons*. Dulles: Brassey's.

Pidgeon, Nick, Karen Henwood and Peter Simmons (2008). *Living with Nuclear Power in Britain: A Mixed-Methods Study. Social Contexts and Responses to Risk (SCARR) Network*, September. Available at: www.kent.ac.uk/scarr/SCARRNuclearReportPidgeonetalFINAL3.pdf.

Pilat, Joseph F. (ed.) (2007). *Atoms for Peace: A Future After Fifty Years?* Washington, DC: Woodrow Wilson Center Press.

Pomper, Miles (2009). 'The Russian Nuclear Industry: Status and Prospects', *Nuclear Energy Futures Paper*, No. 3, January. Available at: www.cigionline.org/publications/2009/1/russian-nuclear-industry-status-and-prospects.

Pomper, Miles (2010). 'US International Nuclear Energy Policy: Change and Continuity', *Nuclear Energy Futures Paper*, No. 10, December. Available at: www.cigionline.org.

Poulson, Kevin (2009). 'Nuclear Plants Cautiously Phase Out Dial-up Modems', *Wired*, 12 October. Available at: www.wired.com/threatlevel/2009/10/nuke_modems/.

Public Opinion Analysis Sector, European Commission (2008). 'Attitudes Towards Radioactive Waste Management', *Special Eurobarometer 297/Wave 69.1 – TNS Opinion &*

Social, June. Available at: http://ec.europa.eu/public_opinion/archives/ebs/ebs_297_en. pdf.

Quester, George H. (1985). 'Taiwan', in Jozef Goldblat (ed.) *Non-Proliferation: the Why and Wherefore*, Stockholm International Peace Research Institute (SIPRI). London: Taylor & Francis, pp. 227–234.

Ramana, M.V. (2009). 'The Indian Nuclear Industry: Status and Prospects', *Nuclear Energy Futures Paper*, No. 9, December. Available at: www.cigionline.org/publications/2009/12/indian-nuclear-industry-status-and-prospects.

Ramavarman T. (2009). 'UAE-US Nuclear Deal Becomes a Reality', *Khaleej Times*, 17 November. Available at: www.khaleejtimes.com/DisplayArticle08.asp?xfile=data/theuae/2009/November/theuae_November529.xml&section=theuae.

Ramsey, Charles B. and Mohamed Modarres (1998). *Commercial Nuclear Power: Assuring Safety for the Future*. New York: John Wiley & Sons, Inc.

Rautenbach, Johan, Wolfram Tonhauser and Anthony Wetherall (2006). 'Overview of the International Legal Framework Governing the Safe and Peaceful Uses of Nuclear Energy–Some Practical Steps', in *International Nuclear Law in the Post-Chernobyl Period*, Joint Report by OECD/NEA and IAEA.

Research Council of Norway (2008). 'Thorium Committee Submits Report: Neither Dismisses nor Embraces Thorium Fuel', April. Available at: www.forskningsradet.no/servlet/Satellite?c=GenerellArtikkel&cid=1203528274171.

Reuters (2007). 'UN's ElBaradei Discusses Jordan's Nuclear Plans', 15 April. Available at: www.reuters.com/article/topNews/idUSL1557493820070415.

Reuters (2009). 'EDF, GDF Suez get OK for French EPR Nuclear Site', 29 January. Available at: http://uk.reuters.com/article/idUKLT28869320090129.

Romm, Joseph (2008). 'The Self-Limiting Future of Nuclear Power'. Center for American Progress Action Fund, Washington, DC, 2 June. Available at: www.americanprogressaction.org/issues/2008/nuclear_power_report.html.

Ryan, Margaret L. (2008). 'US Units Shine, But World Nuclear Generation Lags in 2007', *Nucleonics Week*, 14 February.

Salama, Samir (2008). 'UAE will be First Gulf State to Develop Civilian Nuclear Power', *Gulf News*, 24 March. Available at: http://gulfnews.com/news/gulf/uae/government/uae-will-be-first-gulf-state-to-develop-civilian-nuclear-power-1.92883.

Saraf, Sunil (2008). 'Uranium Shortage Dented Output of India's Power Reactors Last Year', *NuclearFuel*, 2 June.

Scheinman, Lawrence (ed.) (2008). *Implementing Resolution 1540: The Role of Regional Organizations*. New York: UNIDIR.

Schewe, Phillip F. (2007). *The Grid.* Washington, DC: National Academies Press.

Schmitt, Eric (2008). 'Preparing for the Nuclear Power Renaissance', *Capgemini.* Available at: www.capgemini.com/insights-and-resources/by-publication/preparing_for_the_nuclear_power_renaissance/.

Schneider, Mycle (2008a). 'The Reality of France's Aggressive Nuclear Power Push', *Bulletin of the Atomic Scientists*, 3 June. Available at: www.thebulletin.org/web-edition/op-eds/the-reality-of-frances-aggressive-nuclear-power-push.

Schneider, Mycle (2008b). 'Nuclear Power in France: Beyond the Myth', December. Available at: www.greens-efa.org/cms/topics/dokbin/258/258614.mythbuster@en.pdf.

Schneider, Mycle (2009). 'Nuclear Power in France – Trouble Lurking Behind the Glitter', in L. Mez, M. Schneider and S. Thomas (eds) *International Perspectives on Energy Policy and the Role of Nuclear Power*. Brentwood, UK: Multi-Science Publishing.

Schneider, Mycle and Anthony Froggatt (2008). 'The World Nuclear Industry Status

Report 2007'. Brussels, London and Paris: The Greens/European Free Alliance. January.

Schneider, Mycle, Steve Thomas, Antony Froggatt and Doug Koplow (2009). 'The World Nuclear Industry Status Report 2009'. Commissioned by the German Federal Ministry of Environment, Nature Conservation and Reactor. Available at: www.bmu.de/files/english/pdf/application/pdf/welt_statusbericht_atomindustrie_0908_en_bf.pdf.

Schneidmiller, Chris (2008). 'International Agreement Needed on Nuclear Security Standards, NNSA Chief Says', *Global Security Newswire*, Nuclear Threat Initiative, 18 September. Available at: www.nti.org/d_newswire/issues/2008_9_18.html.

Schwartz, Julia A. (2006). 'International Nuclear Third Party Liability Law: The Response to Chernobyl', in *International Nuclear Law in the Post-Chernobyl Period*, Joint Report by the OECD/NEA and IAEA.

Sermage, Jérôme (2009). 'Carriage of Concentrates', *NEI Magazine*, December.

Shakir, Mohamed (2008). Presentation given at the 'Nuclear Energy Futures Project', conference hosted by the Centre for International Governance Innovation, 5–7 November. Waterloo, Canada.

Smith, Aileen Mioko (2007). 'The Failures of Japan's Nuclear Fuel Cycle Program 1956–2007'. Japan: Green Action Kyoto, May. Available at: www.citizen.org/documents/AileenMiokoSmithPresentation.pdf.

Smith, Brice (2006). *Insurmountable Risks: The Dangers of Using Nuclear Power to Combat Global Climate Change*. A Report of the Institute for Energy and Environmental Research (IEER). Muskegon, Michigan: IEER Press.

Smith, Michelle M. and Charles D. Ferguson. (2008). 'France's Nuclear Diplomacy', *International Herald Tribune*, 11 March.

Socolow, Robert H. and Alexander Glaser (2009). 'Balancing Risks: Nuclear Energy and Climate Change', *Daedalus*, Vol. 138, No. 4, pp. 31–44.

Sokolski, Henry (2007). *Falling Behind: International Scrutiny of the Peaceful Atom*. Washington, DC: Nonproliferation Policy Education Center.

Sokolski, Henry (ed.) (2008). *Falling Behind: International Scrutiny of the Peaceful Atom, Strategic Studies Institute*. Carlisle, PA: US Army War College.

Spector, Leonard (1988). *The Undeclared Bomb*. Cambridge, MA: Ballinger Publishing Company.

Spotts, Peter N. (2010). 'Obama Advances Nuclear Resurgence with US Loan Guarantee', *Christian Science Monitor*, 16 February. Available at: www.csmonitor.com/USA/2010/0216/Obama-advances-nuclear-resurgence-with-US-loan-guarantees.

Squassoni, Sharon (2009a). *Nuclear Energy: Rebirth or Resuscitation?* Washington, DC: Carnegie Endowment for International Peace.

Squassoni, Sharon (2009b). 'The US Nuclear Industry: Current Status and Prospects under the Obama Administration', *Nuclear Energy Futures Paper*, No. 7, November. Available at: www.cigionline.org/sites/default/files/Nuclear_Energy_7_0.pdf.

Stablum, Anna (2008). 'Shipping Bottlenecks May Halt Nuclear Renaissance', Reuters, 27 February.

Standard & Poor's (2009). *Sovereign Credit Rating*, August. Available at: http://www.2.standardandpoors.com/portal/site/sp/en/ca/page.topic/ratings_sov/2,1,8,0,0,0,0,0,0,0,0,0,0,0,0.html.

Stanley Foundation (2009). 'Implementing UNSCR 1540: Next Steps Towards Preventing WMD Terrorism', *Policy Memo*, 18 December. Available at: www.stanleyfoundation.org/publications/policy_memo/ImplementUNSCR15401209PM.pdf.

Starr, Barbara (2007). 'Air Force Investigates Mistaken Transport of Nuclear Warheads',

CNN, 6 September. Available at: www.cnn.com/2007/US/09/05/loose.nukes/index. html.

Statens Strålevern (Norwegian Radiation Protection Authority) (2008). 'Floating Nuclear Power Plants and Associated Technologies in the Northern Areas', *Strålevern Rapport No. 15*. Oslo.

Steed, Roger G. (2007). *Nuclear Power in Canada and Beyond*. Renfrew: General Store Publishing House.

Stellfox, David (2007). 'Latin American, Spanish Regulators to Add Reactor Safety to Forum', *Inside NRC*, 23 July, p. 16.

Stone, Time (2008). 'Nuclear Regulatory Review: Summary Recommendations', Office for Nuclear Development, HM Government (UK). December. Available at: www.berr. gov.uk/files/file49848.pdf.

Sub-Committee on Energy and Environmental Security (2009). 'The Nuclear Renaissance', Draft Report. Brussels: NATO Parliamentary Assembly, 17 March.

Tarvainen, Matti (2009). 'Unfair Trade: Nuclear Trade Analysis May Provide Early Indications of Proliferation', *IAEA Bulletin*, Vol. 50, No. 2, pp. 61–63.

Tetley, Mark (2006). 'Revised Paris and Vienna Nuclear Liability Conventions – Challenges for Nuclear Insurers', *Nuclear Law Bulletin*, No. 77.

Thomas, Stephen, Peter Bradford, Antony Froggatt and David Milborrow (2007). 'The Economics of Nuclear Power: Technology, Economics, Alternatives & Case Studies'. Greenpeace International, Amsterdam, Annex A. Available at: www.greenpeace.org. uk/files/pdfs/nuclear/nuclear-economics-report.pdf.

Thomas, Steve (2005). 'The Economics of Nuclear Power: Analysis of Recent Studies'. Public Services International Research Unit (PSIRU), University of Greenwich, July. Available at: www.psiru.org/reports/2005–09-E-Nuclear.pdf.

Toronto Star (2009). '26B Cost Killed Nuclear Bid', 14 July. Available at: www.thestar. com/article/665644.

Transparency International (2009). 'Corruption Perceptions Index 2009'. Available at: www.transparency.org/policy_research/surveys_indices/cpi/2009.

Trosman, Greg (2009). 'Nuclear Safety and Energy Security', in S. Apikyan and D.J. Diamond (eds) *Nuclear Power and Energy Security, NATO Science for Peace and Security Series – B: Physics and Biophysics*. Dordrecht: Springer, pp. 63–67.

Tucker, William (2008). *Terrestrial Energy: How Nuclear Power Will Lead the Green Revolution and End America's Energy Odyssey*. Savage: Bartleby Press.

Tucker, William (2009) 'There is No Such Thing as Nuclear Waste', *The Wall Street Journal*, 13 March.

TVO (2009). 'Start-up of Olkiluoto 3 Nuclear Power Plant May be Postponed Further', Press Release, Teollisuuden Voima Oyj, 15 October. Available at: www.tvo.fi/www/ page/3266/.

UN (1978). 'Resolution Adopted on the Report of the Ad hoc Committee of the Tenth Special Session', UN document A/RES/S-10/2, 30 June.

UN (2000). 'Final Document of the 2000 Review Conference of the Parties to the Treaty of the Non-Proliferation of Nuclear Weapons', UN document NPT/CONF.2000/28, New York.

UN (2004). 'Report of the Secretary-General's High-Level Panel on Threats, Challenges and Change', United Nations document A/59/565, New York, 2 December.

UN (2005). 'International Convention for the Suppression of Acts of Nuclear Terrorism', *United Nations Treaty Series*, 13 April. Available at: http://untreaty.un.org/cod/avl/ha/ icsant/icsant.html (accessed 27 November 2009).

UNIDIR (2003). *Coming to Terms with Security: A Handbook on Verification and Compliance, UNIDIR and VERTIC*. Geneva: United Nations.

US Department of Energy (DOE) (2009). 'Expected New Nuclear Power Plant Applications', Nuclear Regulatory Commission, 28 September. Available at: www.nrc.gov/reactors/new-reactors/new-licensing-files/expected-new-rx-applications.pdf.

US Department of State (2009a). 'Joint Co-Chair Statement at 2009 Plenary Meeting', Bureau of International Security and Nonproliferation (ISN), Washington, DC, 16 June. Available at: www.state.gov/t/isn/rls/fs/125325.htm (accessed 27 November 2009).

US Department of State (2009b). 'Participants', Proliferation Security Initiative, Bureau of International Security and Nonproliferation, US Department of State, Washington, DC, 27 May. Available at: www.state.gov/t/isn/c10390.htm.

US Department of State (2009c). 'The Global Initiative to Combat Nuclear Terrorism', Bureau of International Security and Nonproliferation (ISN). Available at: www.state.gov/t/isn/c18406.htm (accessed 27 November 2009).

US Government Printing Office (GPO) (2009a). 'Foreign Exchange Rates, 1985–2008' (Table B-110), *Economic Report of the President*, Washington, DC.

US Government Printing Office (GPO) (2009b). 'Gross Domestic Product and Deflators Used in the Historical Tables: 1940–2014' (Table 10.1), *Budget of the United States Government (FY 2010)*, Washington, DC.

US Office of Management and Budget (2006). 'Contributions to the IAEA', The White House, Washington, DC. Available at: www.whitehouse.gov/omb/expectmore/summary/10004639.2006.html.

US Presidential Committee of Advisors on Science and Technology (1995). *Fusion Review Panel, US Program of Fusion Energy Research and Development*, Washington, DC.

Van Hong, Le and Hoang Anh Tuan (2004). 'Preparation Studies for Introduction of Nuclear Power to Vietnam', Vietnam Atomic Energy Commission, Hanoi. Available at: www.vaec.gov.vn/Userfiles/file/NP-Vietnam.pdf.

Vandenbosch, Robert and Susanne E Vandenbosch (2007). *Nuclear Waste Stalemate: Political and Scientific Controversies*. Salt Lake City: University of Utah Press.

von Hippel, Frank (2008). 'Nuclear Fuel Recycling: More Trouble than it's Worth', *Scientific American*, Vol. 28, April.

Wald, Matthew L. 'U.S. Supports New Nuclear Reactors in Georgia', *New York Times*, 16 February. Available at: www.nytimes.com/2010/02/17/business/energy-environment/17nukes.html.

WANO (2007). 'Nuclear Safety Never Better – But Still More to Do', address by WANO Chairman William Cavanagh III, to the WANO Biennial General Meeting, Chicago, 25 September.

WANO (2009). '2008 Performance Indicators', *WANO*, London, April. Available at: www.wano.org.uk/PerformanceIndicators/PI_Trifold/PI_2008_TriFold.pdf.

WBCSD (2008). 'Power to Change: A Business Contribution to a Low-Carbon Electricity Future', Geneva.

Weil, Jenny (2007). 'WANO Warns Safety Lapse Anywhere Could Halt "Nuclear Renaissance"', *Nucleonics Week*, 27 September, pp. 1 and 14.

Weil, Jenny (2009a). 'Klein: Excellent Operations Can Lead to Problems', *Nucleonics Week*, 19 March, p. 1.

Weil, Jenny (2009b). 'NEI's New President Sees Scaled-Down Expansion Goal', *Nucleonics Week*, 12 March.

Weil, Jenny (2009c). 'NRC's Updated Security Rule Takes Effect in May', *Nucleonics Week, Platts*, 2 April, pp. 1–3.

Weinberg, Alvin M., Irving Spiewak, Doan L. Phung and Robert S. Livingston (1985). 'A Second Nuclear Era: A Nuclear Renaissance', *Energy*.

WENRA (2009). 'General Presentation', Western European Nuclear Regulators' Association (WENRA), August. Available at: www.wenra.org/dynamaster/file_archive/09112 3/2bc4835423c9024eff8542c20993085a/WENRA%20%2d%20General%20Presentation.pdf.

Whitlock, Jeremy J. (2000). 'The Evolution of CANDU Fuel Cycles and Their Potential Contribution to World Peace', *Atomic Energy of Canada Limited (AECL)*, April. Available at: www.nuclearfaq.ca/brat_fuel.htm.

Windsor, Lindsay and Carol Kessler (2007). 'Technical and Political Assessment of Peaceful Nuclear Power Program Prospects in North Africa and the Middle East', Pacific Northwest Center for Global Security, September.

Winner, Andrew C. (2005). 'The Proliferation Security Initiative: The New Face of Interdiction', *The Washington Quarterly*, Vol. 28, No. 2, p. 129.

WINS (2009). World Institute for Nuclear Security (website). Available at: www.wins.org (accessed 26 November 2009).

WMD Commission (2006). 'Weapons of Terror: Freeing the World of Nuclear, Biological and Chemical Arms', Stockholm: Weapons of Mass Destruction Commission. Available at: www.wmdcommission.org/files/Weapons_of_Terror.pdf.

World Bank (2003). 'Pages from World Bank History: Loan for Nuclear Power', 22 August. Available at: http://web.worldbank.org/WBSITE/EXTERNAL/EXTABOUTUS/ EXTARCHIVES/0,,contentMDK:20125474~pagePK:36726~piPK:36092~theSite PK:29506,00.html.

World Bank (2009). 'World wide Governance Indicators: 1996–2008'. Available at: http:// info.worldbank.org/governance/wgi/index.asp.

World Bank Environment Department (1994). *Environmental Assessment Sourcebook*. Vol. III, World Bank Technical Paper No. 154, April.

World Business Council for Sustainable Development (WBCSD) (2008). 'Power to Change: A Business Contribution to a Low-Carbon Electricity Future', Geneva.

World Information Service on Energy (2010). 'Jordan: Status of Uranium Mining', WISE Uranium Project, 6 January. Available at: www.wise-uranium.org/upasi. html#JOGEN.

World Nuclear Association (WNA) (2006). 'Ensuring Security of Supply in the International Nuclear Fuel Cycle', WNA Report. Available at: www.world-nuclear.org/reference/pdf/security.pdf.

World Nuclear Association (WNA) (2008a). 'Benefits Gained Through International Harmonization of Nuclear Safety Standards for Reactor Designs', Discussion paper, WNA, London, January.

World Nuclear Association (WNA) (2008b). 'Supply of Uranium', June. Available at: www.world-nuclear.org/info/inf75.html.

World Nuclear Association (WNA) (2008c). 'The New Economics of Nuclear Power', WNA Report, November. Available at: www.world-nuclear.org/reference/pdf/economics.pdf.

World Nuclear Association (WNA) (2009a). 'Advanced Nuclear Power Reactors'. Available at: www.world-nuclear.org/info/inf08.html.

World Nuclear Association (WNA) (2009b). *Nuclear Century Outlook*. Available at: www.world-nuclear.org/outlook/nuclear_century_outlook.html.

World Nuclear Association (WNA) (2009c). 'Nuclear Power in Ukraine', World Nuclear Association, December. Available at: http://world-nuclear.org/info/inf46.html.

World Nuclear Association (WNA) (2009d). 'Structuring Nuclear Projects for Success: an Analytical Framework', London. Available at: www.world-nuclear.org/uploaded-Files/org/reference/pdf/EconomicsReport.pdf.

World Nuclear Association (WNA) (2009e). 'Thorium', August. Available at: www.world-nuclear.org/info/inf62.html.

World Nuclear Association (WNA) (2010a). 'China's Nuclear Fuel Cycle', May. Available at: www.world-nuclear.org/info/inf63b_china_nuclearfuelcycle.html#R_and_D).

World Nuclear Association (WNA) (2010b). 'Emerging Nuclear Energy Countries', 25 May. Available at: www.world-nuclear.org/info/inf102.html.

World Nuclear Association (WNA) (2010c). 'Nuclear Power in South Korea', April. Available at: www.world-nuclear.org/info/inf81.html.

World Nuclear Association (WNA) (2010d). 'Nuclear Power in Switzerland', January. Available at: www.world-nuclear.org/info/inf86.html.

World Nuclear Association (WNA) (2010e). 'World Nuclear Power Reactors and Uranium Requirements', 3 May. Available at: www.world-nuclear.org/info/reactors.html.

World Nuclear News (WNN) (2007). 'UK Poll Shows Support for Nuclear, but Confusion on Issues', 5 December. Available at: www.world-nuclear-news.org/newsarticle.aspx?id=14512&LangType=2057.

World Nuclear News (WNN) (2008a). 'Crossroads for Yucca Mountain', 19 December. Available at: www.world-nuclear-news.org/WR_Crossroads_for_Yucca_Mountain_1012121.html.

World Nuclear News (WNN) (2008b). 'GAO: Not All US Reactors Meet Fire Regulations', 1 July. Available at: www.world-nuclear-news.org/RS-Not_all_US_reactors_meeting_fire_regulations_says_GAO-0107084.html.

World Nuclear News (WNN) (2008c). 'IAEA: Nuclear Capacity Could Double by 2030', 12 September. Available at: www.world-nuclear-news.org/NP-IAEA-Nuclear_capacity_could_double_by_2030–1209084.html.

World Nuclear News (WNN) (2008d). 'National Skills Academy Approved for UK Nuclear Industry', 18 January. Available at: www.world-nuclear-news.org/newsarticle.aspx?id=14106.

World Nuclear News (WNN) (2008e). 'New Date for Atucha 2', 18 August. Available at: www.world-nuclear-news.org/IT-Atucha_II_to_start_up_by_October_2010–1808085.html.

World Nuclear News (WNN) (2008f). 'Nuclear Power is 'Vital' to Britain', 11 January. Available at: www.speroforum.com/a/13531/Nuclear-power-is-vital-to-Britain.

World Nuclear News (WNN) (2008g). 'Opinion Favors Nuclear', 29 April. Available at: www.world-nuclear-news.org/NP-Opinion_favours_nuclear_2904089.html.

World Nuclear News (WNN) (2008h). 'Poll: Two Thirds of Americans Back New Nuclear', 9 June.

World Nuclear News (WNN) (2008i). 'Positive Thinking in Italy, Canada, and Poland', 7 August. Available at: www.world-nuclear-news.org/NP-Positive_thinking_in_Italy_Canada_and_Poland-0708087.html.

World Nuclear News (WNN) (2008j). 'UAE Launches Nuclear Policy to Meet Energy Demand', 14 July. Available at: www.world-nuclear-news.org/NP-UAE_launches_nuclear_policy_to_meet_energy_demand-1407086.html.

World Nuclear News (WNN) (2008k). 'UK Opinion Swinging Towards Nuclear', 25 June.

Available at: www.world-nuclear-news.org/NP-UK_opinion_swinging_towards_nuclear-2506087.html.

World Nuclear News (WNN) (2008l). 'Vietnam Passes Law on Nuclear Energy', 4 June. Available at: www.world-nuclear-news.org/IT-Vietnam_passes_law_on_nuclear_energy-0406085.html.

World Nuclear News (WNN) (2009a). 'China to Set Even Higher Nuclear Targets', 1 June. Available at: www.world-nuclear-news.org/NN_China_to_set_even_higher_nuclear_targets_0106091.html.

World Nuclear News (WNN) (2009b). 'Construction at Sanmen within a Month', 2 March.

World Nuclear News (WNN) (2009c). 'Full Power for Japan's First MOX Burner', 3 December. Available at: www.world-nuclear-news.org/IT-Full_power_for_Japans_first_MOX_burner-0312097.html.

World Nuclear News (WNN) (2009d). 'More Shield Work on AP1000', 16 October. Availableat:www.world-nuclear-news.org/RS_More_shield_work_on_AP1000_1610091.html.

World Nuclear News (WNN) (2009e). 'NEI Proposes Legislation for Nuclear Growth', 28 October. Available at: www.world-nuclear-news.org/NP-NEI_proposes_legislation_to_support_nuclear_growth-2810095.html.

World Nuclear News (WNN) (2009f). 'Obama Dumps Yucca Mountain', 27 February. Available at: www.world-nuclear-news.org/newsarticle.aspx?id=24743.

World Nuclear News (WNN) (2009g). 'Olkiluoto 3 Losses to Reach €1.7 Billion', 26 February. Available at: www.world-nuclear-news.org/newsarticle.aspx?id=24732.

World Nuclear News (WNN) (2009h). 'Thorium-fuelled Exports Coming From India', 17 September. Available at: www.world-nuclear-news.org/NP_Thorium_exports_coming_from_India_1709091.html.

World Nuclear News (WNN) (2009i). 'UAE Picks Korea as Nuclear Partner', 29 December. Available at: www.world-nuclear-news.org/NN_UAE_picks_Korea_as_nuclear_partner_2812091.html.

World Nuclear News (WNN) (2010a). 'Changes for International Nuclear Safety', 28 April. Available at: www.world-nuclear-news.org/RS_Changes_for_international_nuclear_safety_2804101.html.

World Nuclear News (WNN) (2010b). 'Indian Nuclear Joint Venture Agreement Signed', 27 April. Available at: www.world-nuclear-news.org/C-Indian_nuclear_joint_venture_agreement_signed-2704105.html.

World Nuclear University (WNU) (2010). 'What is the WNU?'. Available at: www.world-nuclear-university.org/about.aspx?id=15036.

Xinhua (2008). 'China, Jordan Sign Power Plant Deal', 23 September. Available at: http://news.xinhuanet.com/english/2008–09/23/content_10099137.htm.

Yamaguchi, Mari (2010). 'Japan Reactor Reopens After 1995 Accident', The Associated Press. Available at: www.google.com/hostednews/ap/article/ALeqM5jnS4EjX7kdRX-slv4gJYhuM1mcckwD9FH4H7O0.

Yudin, Yury (2009). *Multilateralization of the Nuclear Fuel Cycle: Assessing the Existing Proposals*. Geneva: UNIDIR.

Index

Note: locators in **bold** type indicate figures or illustrations, those in *italics* indicate tables.